W9-DET-931

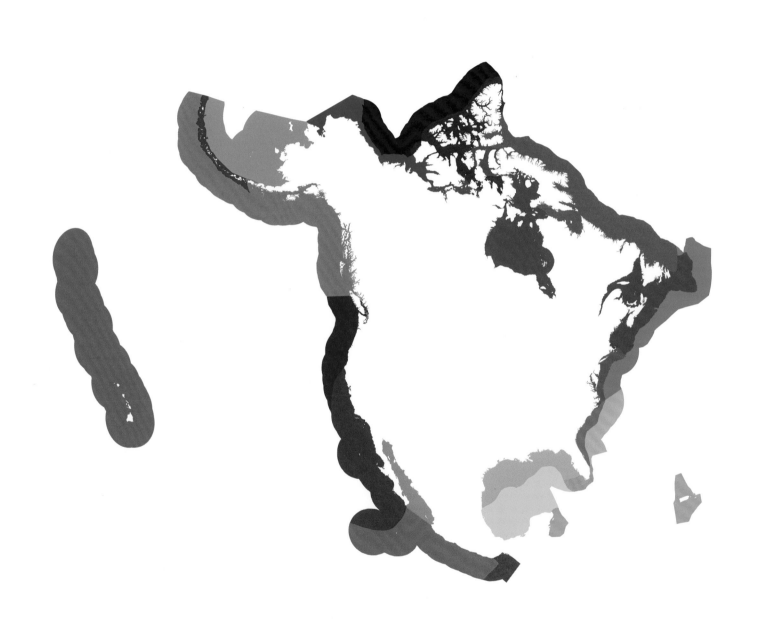

MARINE ECOREGIONS
of North America

This report was written for the Secretariat of the Commission for Environmental Cooperation as a collaborative effort with the following authors of the Marine Ecoregions Project Team. Its publication has been made possible thanks to the generous support of its copublishers. Numerous individuals have contributed to the completion of this work (please see the acknowledgments section for additional information). The information contained herein is the responsibility of the authors and does not necessarily reflect the views of the CEC, or the governments of Canada, Mexico or the United States of America.

Authors

Tara A. C. Wilkinson, McGill University, Commission for Environmental Cooperation (CEC)

Ed Wiken, Wildlife Habitat Canada (WHC)

Juan Bezaury Creel, The Nature Conservancy Mexico (TNC Mexico)

Thomas F. Hourigan, National Oceanic and Atmospheric Administration (NOAA)

Tundi Agardy, Sound Seas

Hans Herrmann, Commission for Environmental Cooperation (CEC)

Lisa Janishevski, Commission for Environmental Cooperation (CEC)

Chris Madden, Natureserve

Lance Morgan, Marine Conservation Biology Institute (MCBI)

Moreno Padilla, Canadian Council on Ecological Areas (CCEA)

Co-Publishers

Comisión Nacional de Áreas Naturales Protegidas

Comisión Nacional para el Conocimiento y Uso de la Biodiversidad

National Oceanic and Atmospheric Administration

Parks Canada

Instituto Nacional de Ecología

Reproduction of this publication, in whole or in part and in whatever form, may be done without seeking express authorization of the CEC, provided that the reproduced material is to be used for educational or nonprofit purposes and that it cites the source of the original. The CEC would appreciate receiving a copy of all publications or material that utilize this work as a source.

Published by the Communications Department of the CEC Secretariat.

© Commission for Environmental Cooperation, 2009

Cite as:

Wilkinson T., E. Wiken, J. Bezaury-Creel, T. Hourigan, T. Agardy, H. Herrmann, L. Janishevski, C. Madden, L. Morgan, M. Padilla. 2009. Marine Ecoregions of North America. Commission for Environmental Cooperation. Montreal, Canada. 200 pp.

ISBN 978-2-923358-41-3 (print version); ISBN 978-2-923358-71-0 (electronic version)

Legal deposit – Bibliothèque national du Québec, 2009

Legal deposit – National Library of Canada, 2009

Disponible en Español: ISBN 978-2-923358-42-0 (versión impresa); ISBN 978-2-923358-72-7 (versión electrónica)

Disponible en français: ISBN 978-2-923358-43-7 (version imprimée); ISBN 978-2-923358-73-4 (version électronique)

For more information about this or other publications from the CEC, contact:

Commission for Environmental Cooperation

393, rue St-Jacques Ouest, Bureau 200

Montreal (Quebec) Canada H2Y 1N9

T 514.350.4300 F 514.350.4314

info@cec.org / http://www.cec.org

Mixed Sources

Cert no. SW-COC-001271
© 1996 FSC

FSC

Printed in Canada

MARINE ECOREGIONS
of North America

Commission for Environmental Cooperation
Comisión para la Cooperación Ambiental
Commission de coopération environnementale

Preface

The wealth of North America's marine and coastal waters is unparalleled. These rich storehouses of biodiversity underpin our quality of life, our economies and much of our cultural identity. From the prolific areas of the Grand Banks of Newfoundland to the exceptionally diverse Mesoamerican Coral Reef, Canada, Mexico, and the United States share a vast array of ecosystems—an interconnected continental web of life, dynamic and wondrous. Yet, when one looks at the continent as a whole, this bewildering assemblage of marine life, already threatened by land-based pollution, overfishing, and invasive species, among a wide array of stressors, is now facing new challenges from rapidly changing climatic conditions.

As much of the damage occurs hidden from our view, under the deceptively unchanging blanket of the ocean's surface, North Americans are seeking new means to protect our common natural endowment. Establishing an effective system for linking places in the ocean to ensure biological connectivity, increased resilience, and protection of ecosystem integrity, required a meaningful ecological framework. In 2002 a trinational group of marine scientists and CEC officials met at NOAA's Coastal Services Center in Charleston, South Carolina, to agree on a new, unified, ecological classification for oceanic and coastal regions.

This book is the result of that endeavor: an approach, system of classification, and map attempting to create consistent, standardized and understandable units out of the vastness of the continent's ocean and coastal waters; a system that could be scalable, ecosystem-oriented, and linked to existing maps and classifications. It presents the developing consensus of American, Canadian, and Mexican ecologists, marine biologists, geographers, planners, and managers on what this continent holds in terms of marine biodiversity.

While the division of the oceans and coasts into discrete parcels is admittedly a difficult task and may even seem an artificial construct, naming and describing places is the only way we can begin to grasp the overwhelming complexity that binds them together. By determining how the continental waters may be catalogued as sets of habitats and assemblages of species, we can begin to understand more fully the great richness that our oceans present. Mapping ecosystems provides us a way of looking at our world on a scale we can fathom.

Using descriptive profiles, this book classifies the ocean and coastal regions of North America into 24 marine ecoregions, large masses of water differentiated by oceanographic features and geographically distinct assemblages of species that interact ecologically in ways that are critical for their long-term persistence.

Each chapter briefly describes the key features of each ecoregion—in terms of its physical, oceanographic, and biological characteristics, as well as the impacts we humans have had upon it. Each chapter also enables the reader to have a quick glance at the state of knowledge by means of *Fact Sheets* to be found near the beginning of each section. These *Fact Sheets* summarize geographical, oceanographical, physiological and biological information, such as the size of the region, its sea surface temperature, its primary productivity, and provide a thumbnail description of the region by depth, its key habitats, species at risk and the principal human activities engaged in there. Sometimes categories may not be present, according to their importance for the region, or to the information available. For example, the category of endemic species is quite extensive within the well-studied Gulf of California ecoregion, but omitted for the Arctic Basin, where information on the subject is more sparse. Each profile also contains information on how the region was delineated.

There are also important sections at the back of the book that contain acronyms and abbreviations; a glossary of common conservation and scientific terms used throughout the book; a list of important species, including endemic and invasive species, highlighted in the text, given by common name (English, French and Spanish) and scientific name; a list of related websites; and a reference list. Finally, and for the sake of completeness, the book also contains short descriptions of the distinct ecoregions of the US Pacific Island Territories.

The resultant framework cannot presume to be a complete, all-encompassing product that will be all things to all people. Rather, its goal was to provide a common starting point, a scalable framework to collect and organize information, encourage further cooperation, and be used as a tool to better understand and manage our North American marine ecosystems.

This unprecedented effort to promote a better understanding of our oceans has been possible thanks to the leadership and generous contributions of *Comisión Nacional de Áreas Naturales Protegidas* (Conanp), *Comisión Nacional para el Conocimiento y Uso de la Biodiversidad* (Conabio), and *Instituto Nacional de Ecología* (INE) in Mexico; the National Oceanic and Atmospheric Administration (NOAA) in the United States, and Parks Canada.

Marine Ecoregions of North America represents an early step in the process of coming to terms with the complex and awe-inspiring marine world of North America. We can now point to marine regions and seascapes, connected not only by species and ecological processes but also by our own movements and interactions with the sea and its creatures. With better knowledge of what lies under that watery blanket, together, we can move more confidently to protect it.

Hans Herrmann
CEC

Acknowledgments and Contributors

Charleston Drafting Meeting Participants

Rebecca Allee (National Oceanic and Atmospheric Administration, NOAA), Saúl Álvarez Borrego (Centro de Investigación Científica y de Educación Superior de Ensenada, CICESE), Jeff Ardron (Living Oceans Society, LOS), Juan Bezaury Creel (WWF, Programa México), Luis Eduardo Calderón (CICESE-Ecología), Arturo Carranza Edwards (Universidad Nacional Autónoma de México, UNAM), Kimberly Cohen (NOAA Coastal Services Center), Antonio Díaz de León (Instituto Nacional de Ecología, INE/El Colegio de México), Gilberto Enríquez Hernández (INE), Linda Evers (NatureServe), Zach Ferdaña (The Nature Conservancy of Washington), Gilberto Gaxiola (CICESE), J. B. Heiser (Cornell University), Robert Hélie (Servicio Canadiense de Vida Silvestre), Hans Herrmann (CCA), Andrew Hulin (NOAA Coastal Services Center), Lacy Johnson (NOAA Coastal Services Center), Brenda Konar (Fish and Marine Science Faculty, University of Alaska), Héctor Alfonso Licón González (DICTUS/ Universidad de Sonora), Jon Lien (Memorial University, Newfoundland), Christopher Madden (NatureServe), Francine Mercier (Parks Canada), Claude Mondor (Parks Canada), Moreno Padilla (Wildlife Habitat Canada), Jeffrey Payne (NOAA Coastal Services Center), Heidi Reck Siek (NOAA Coastal Services Center), John Roff (Acadia University), Scott Rutzmoser (NOAA Coastal Services Center), Karen Schmidt (CEC), Juan J. Schmitter Soto (El Colegio de la Frontera Sur, Ecosur), Hamilton Smillie (NOAA Coastal Services Center), Rob Solomon (NatureServe), Margarito Tapia García (Universidad Autónoma Metropolitana, UAMIztapalapa), Kate Thomas (California State University, Monterey Bay), Carlos Valdés (CEC), Ed Wiken (Wildlife Habitat Canada/Canadian Council on Ecological Areas), Tara A. C. Wilkinson (CEC), Alejandro Yáñez Arancibia (INE), Mark Zacharias (California State University Channel Islands, previously at the Ministry of Sustainable Resources Management, Province of British Columbia).

Technical Facilitators for the Charleston Workshop

George Dias (Applied Geomatics Research Group, Annapolis Valley Campus), Daniel Asher Hackett (McGill University), Thomas Meredith (McGill University).

Peer Reviewers

Saúl Álvarez Borrego (Centro de Investigación Científica y de Educación Superior de Ensenada, CICESE), Jeff Ardron (Living Oceans Society, LOS), Peter Auster (University of Connecticut), Doug Ballam (Canadian Council on Ecological Areas, CCEA), Mike Beck (The Nature Conservancy, TNC), Luis Eduardo Calderón (CICESE-Ecología), Arturo Carranza Edwards (Universidad Nacional Autónoma de México, UNAM), Peter J. Celone (National Oceanic and Atmospheric Administration, NOAA), Mike Dunn (Nature-Ed Services), Gilberto Enríquez Hernández (Instituto Nacional de Ecología, INE), Peter Etnoyer (Marine Conservation Biology Institute, MCBI), Glenn Ford, Mark Mallory (Canadian Wildlife Service, CWS), Francine Mercier (Parks Canada), Claude Mondor (Parks Canada), Lance Morgan (MCBI), G. Carleton Ray (University of Virginia), John C. Roff (Acadia University), Carl Schoch (Prince William Sound Science Center, previously at Kachemak Bay Research Reserve), Ken Sherman (NOAA), Juan J. Schmitter Soto (El Colegio de la Frontera Sur, Ecosur), Hamilton Smillie (NOAA), Kathleen Sullivan Sealey (University of Miami), Margarito Tapia García (Universidad Autónoma Metropolitana, UAM-Iztapalapa), Tony Turner (A.M. Turner and Associates), Mark Zacharias (California State University Channel Islands, previously at the Ministry of Sustainable Resources Management, Province of British Columbia).

Experts Consulted

Alfonso Aguirre Muñoz, Grupo de Ecología y Conservación de Islas, A.C., Martha Basurto Origel, Centro Regional de Investigación Pesquera (CRIP) Puerto Morelos, Bruce Amos (IUCN), Natalie Ban (CPAWS), Therese Beaudet (IUCN), Tom Beechey (Ontario Ministry of Natural Resources, Onatario Parks), James Birtch (Parks Canada), Dan Brumbaugh (American Museum of Natural History and NOAA), Tomás Camarena Luhrs, Environmental Defense, Real Carpentier (Ministere de l'Environment du Quebec), Dick Carson (DFO), Dan Chambers (Alberta Environmental Protection, Natural Resources Services), Flavio Chazaro (Conanp), Doug Chiperzak (Inuvialuit Settlement Region Oceans Program), Jean Cinq-Mars (WHC), Gilberto Cintron (US Fish and Wildlife Service), Mary Jean Comfort (DFO), John Crump (Canadian Arctic Resources Committee), Rosemary Curley (Department of Environmental Resources, Government of P.E.I.), Charles Ehler (NOAA), Ernesto Enkerlin (Conanp), Exequiel Ezcurra (INE), Jean Gagnon (Service des aires protégées, Direction de la conservation e du patrimoine écologique), Gerardo García Beltrán, Amigos de Sian Ka´an, A.C., Mart R. Gross (University of Toronto), Denny Grossman (NatureServe), David Gutiérrez Carbonell, Conanp, Helios Hernandez (Parks and Natural Areas Branch, Department of Natural Resources, Government of Manitoba), Don Howes (Landuse Coordination Office, Government of British Columbia), David Hyrenbach (PRBO), Glen Jamieson (DFO), Sabine Jessen (Canadian Parks and Wilderness Society), Marty King (WWF Canada), Tiina Kurvitz (UNEP-GRID), Jennifer Lash (Living Oceans Society), Claudia Padilla Souza, CRIP Puerto Morelos, Mário Lara Pérez-Soto, Conanp, Josh Laughren (WWF Canada), Nik Lopoukhine (Parks Canada), Rosa María Loreto Viruel, Amigos de Sian Ka´an, A.C., Ian Marshall, Jack A. Mathias (DFO), Don McAllister (Ocean Voice International), Kevin McCormick (CWS), John Meikle (Government of Yukon), Lynne Mersfelder (NOAA), Mark Monaco (NOAA), Harold Moore (GeoInsight Corp.), Ken Morgan (CWS), Ken Morrison (BC Parks), Chris Morry (IUCN), Sebastian Oosenbrug (Wildlife and Fisheries Division, Dept. of Renewable Resources, Government of Northwest Territories), Simona Perry (NOAA), Francine Proulx (IUCN Canada), Oscar Ramírez Flores, Dirección General de Vida Silvestre, Semarnat, Kathryn Ries (NOAA), Cheri Recchia (The Ocean Conservancy), Lorenzo Rojas (INE), Doug Ryan (US Fish Wildlife and Service), Glen Ryan (Parks and Natural Areas Division, Department of Tourism, Culture, and Recreation, Government of Newfoundland), Enric Sala (Scripps Institute of Oceanography), Dale Smith (Parks and Recreation Division, Department of Natural Resources, Government of Nova Scotia), Jennifer Smith (WWF Canada), Leigh Warren (Canadian Wildlife Service), Lani Watson (NOAA), John P. Vandall (Resource Management and Protection, Government of Saskatchewan), Herb Vandermeulen (DFO), Liette Vasseur (University of Moncton), Doug Yurick (Parks Canada), Charles Wahle (NOAA), Darren Williams (DFO), Larry Wolfe, Vincent Zelazny (Department of Natural Resources and Energy, Government of New Brunswick)

Other Experts Consulted

Rebecca Allee, Dave Canny, Flavio Cházaro, Kimberly Cohen, Antonio Díaz de León, Michael Dunn, Linda Evers, Dan Farrow, Zach Ferdaña, Lloyd T. Findley, Gilberto Gaxiola, Mike Goard, Robert Hélie, J. B. Heiser, Don Howes, Andrew Hulin, Lacy Johnson, Brenda Konar, Tiina Kurvits, Tony Lavoi, Héctor Alfonso Licón González, Jon Lien, Claude Mondor, Percy Pacheco, Jeffrey Payne, Scott Rutzmoser, Heidi Reck Siek, Hamilton Smiley, Kate Thomas, Joseph Uravitch, Alejandro Yáñez Arancibia

Informatics and Mapping

Carlos Valdés, Daniel Asher Hackett, Fernando Gutiérrez, John Nick Sanders, Linda Evers; Rob Solomon

Translators

Virginia Aguirre, Juan Bezaury Creel, Marie-Claude Guy, Hans Herrmann, Raymonde Lanthier, Raúl Marcó del Pont, Marina Margarita Molas, Low Pfeng, Renato Rivera, Silvia Ruiz de Chávez

CEC Editorial Team

Jeffrey Stoub, Jacqueline Fortson, Douglas Kirk, Johanne David; Karen Schmidt, Itzel Hernández

Layout and Design

Patricio Robles Gil (Sierra Madre); Gray Fraser

Often called the "sea cow," the gentle and slow moving West Indian manatee inhabits the warmer Atlantic and Gulf of Mexico coastal waters. *Photo:* Doug Perrine/DRK PHOTO

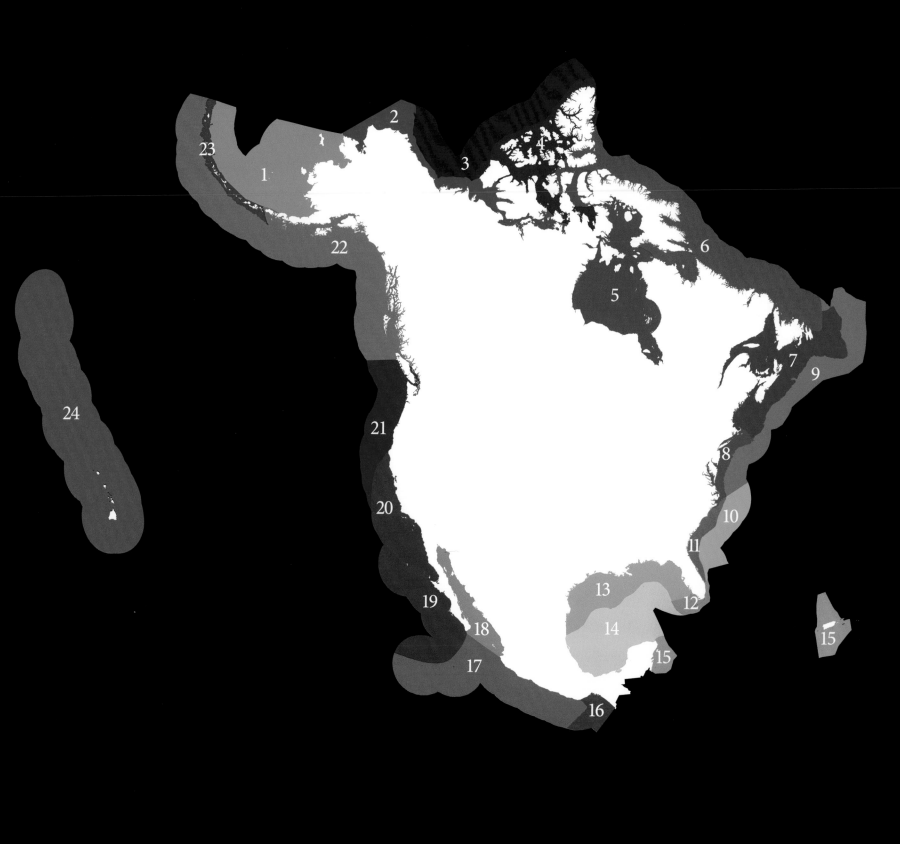

Table of Contents

Green moray eels are sedentary
predators that wait until food comes
to them rather than hunting for it.
Photo: Octavio Aburto

A small pod of Atlantic spotted dolphins, a species that typically dwells in the warmer offshore waters in the Atlantic.   *Photo:* Doug Perrine/DRK PHOTO

"All we do is touched with ocean yet we remain
on the shore of what we know."

Richard Wilbur[1]

1   From http://oceanexplorer.noaa.gov/explorations/deepeast01/logs/sep29/sep29.html

Polar bears are among the more unique ocean/land mammals, and majestic symbols for the northern seas.   *Photo:* Patricio Robles Gil

Introduction

Why map ecological regions in marine areas?

A simple glance at a world map shows how dominant the oceans are on earth—over 70 percent of the surface of the planet is covered by blue; if measured by living space, 99 percent is found in the oceans. Marine ecosystems also hold important roles in our lives—aside from providing major habitats for wild species, they impart a major source of food for the world's inhabitants. Moreover, the oceans' ecosystems directly and indirectly provide employment opportunities for many of the coastal residents. Commercial and recreational fishermen, scuba divers, tourism operators, hoteliers, restauranteurs, pharmaceutical companies, researchers, shipping, and oil and gas companies rely on marine ecosystems to help create and support jobs and sustain our economies. In addition, mangroves and reefs help to buffer the impacts of storms, helping defend coastlines against the erosive forces of waves and storm surges. Marshes and coastal wetlands help filter pollution from land-based sources. All provide essential nutrients and contain critical habitat for a wide variety of marine life (e.g., feeding and spawning areas, nurseries, and migratory corridors). The ocean also has its hand in regulating global hydrological and biochemical cycles as well as climate.

Yet to most, the marine realm remains relatively remote and obscure. Found beneath the seemingly homogeneous and unchanging blue blanket of the water's surface, species, habitats and ecosystems of the ocean are relatively isolated from and figure little in the minds of most people. Only the sea's most charismatic life forms, like whales, sea turtles, polar bears, sea otters and coral reefs seem to garner any attention. Yet compared to their land counterparts, the understanding of these marine inhabitants is limited.

As we are affected by the oceans in much of what we do, so too do we affect this realm and its inhabitants—both directly and indirectly. North America is a continent that is particularly reliant on oceans, and one that strongly affects the condition of its marine ecosystems. Those ecosystems now show signs of serious ecological imbalance, habitat destruction, wildlife impacts and biodiversity loss.[2] The formerly rich, and culturally and economically critical fish stocks, such as those of the Northwest Atlantic, have collapsed, throwing thousands out of work. Toxic algal blooms disrupt the food chain and impact human health throughout our coasts. Our coral reefs are suffering from coral bleaching, emergent coral diseases, sedimentation and algal overgrowth. Just inland the ecological damage we wrought

to areas such as the great Everglades through canal building, agricultural waste, and zealous urbanization is expected to cost over one hundred billion dollars to correct. Eutrophication is also prevalent across our shores—in the Gulf of Mexico an expanding "dead zone" of oxygen-deprived and lifeless water caused in part by river-borne pollutants is stressing our marine ecosystems. And as with all other marine systems, the threat of pollution does not act alone—numerous other pressures, such as bottom trawling that scrapes much of the seafloor's complexity and diversity, strain marine ecosystems simultaneously.

Overfishing, coupled with incomplete information on the habitat and life-history requirements of many species, has depleted numerous stocks—such as cod, abalone, twenty plus species of rockfish, and coastal sharks, to name but a few—and has adversely impacted species that depend on these resources, such as the sea otter. At the same time, some of our most valued areas—our treasured beaches—are periodically closed to swimming as bacterial levels exceed health standards. Additionally, several of our runs of salmon are endangered and our coastal waters are suffering from land-based sources of pollution exacerbated by massive logging operations. Moreover, our coastal species, habitats and ecosystems are being crowded out by invasive species—San Francisco Bay and delta harbor more than 234 invasive species. And even in the Arctic, the impacts of global climate change, bioaccumulation of toxic chemicals—like PCBs (polychlorinated biphenyls) and DDT (dichloro-diphenyl-trichloroethane)—from far off sources, and overexploitation of marine resources are increasingly evident, triggering changes in the abundance and behavior of arctic species (CAFF 2004). All this is taking place in one of the world's wealthiest continent—among nations that pride themselves on commitment to the environment. Our lack of an integrated seascape approach to marine resource management has led to a loss in fisheries production, shoreline stabilization, and natural pollutant cleansing processes—to name but a few natural ecological services provided from coastal environments. We are not allowing natural systems to maintain themselves and do the jobs we need them to do, for our own survival and for the survival of the other species with which we share this planet.

Although there are many dedicated conservation efforts and sustainable development initiatives presently targeting North American marine species, habitats and ecosystems, they generally work independently of each other. Coordinating these efforts within an ecosystem-based perspective could help us prevent species numbers from continuing to dwindle and ecosystem integrity from being further at

---

2   Information from the following paragraph was adapted from Institutional Options for Integrated Management of a North American MPA Network, written for the CEC by T.S. Agardy and L. Wolfe (2002).

risk. Successful conservation of North American seascapes requires studying and dealing with the system as a whole. This "whole" can be thought of perhaps best as a discrete area of land or sea—an ecoregion—characterized by a distinctive climate, ecological features, and natural communities. Managing ecoregions in a wise and sustainable way at North American scale is of course a challenge, requiring cooperative action from all three countries and a great many different sectors of society. But it is a vital task if we want to preserve our natural heritage for future generations.

With this in mind, individuals, agencies, institutions and organizations from Canada, Mexico and the United States—under the umbrella of the Commission for Environmental Cooperation—agreed to undertake the work that would lead to Marine Ecoregions of North America. This project responded to a priority need identified through CEC's *Strategic Plan for North American Cooperation in the Conservation of Biodiversity*. It can serve as a common base to help improve marine knowledge, research design, policy initiatives and management decisions. It can also assist in coordinating and guiding management practices and improving outreach and education efforts for the conservation and sustainable use of our common heritage.

Current day management strategies, such as sustainable living and resource use and biodiversity conservation, must be based on robust, relevant knowledge, capable of supporting a flexible analytical framework. Decisions concerning sustainable development, adaptive management, stress/exposure/response indicators, life-cycle care, and integrated seascape planning and management need an ecosystem approach to help unite diverse ocean management needs and respect varied interests of stakeholders and interest groups. These user requirements can be addressed by a marine framework that is:

Scalable—responding to perspectives and interests that may be regional in nature through to those that are continental and worldwide in scope;

Ecosystem-oriented—based on a range of connected biological, oceanographic and physiographic parameters, and capable of supporting various socio-economic factors; and

Linked—with other marine and terrestrial maps and classification systems and disciplines.

The Marine Ecoregions of North America can help to: 1) support the implementation, development and coordination of national and international mandates, conventions, policies and acts, 2) support varied interests of stakeholders, and 3) provide information to the public, nongovernmental organizations, industries and governments. It is also meant to support research and education, inventorying and monitoring and other planning efforts. It is hoped that the delineated ecoregions will serve as the basis for regional and cooperative steward-

ship and management efforts. They can be used as reference points for periodic assessments of ecosystems and their habitats, species and other environmental components. Finally, the definition of marine ecoregions can also help define representative and critical areas of the marine environment through a network of marine protected areas and special conservation areas—cornerstones of ecosystem-based conservation and sustainable development strategies.

It is hoped that this effort will provide a framework to use and integrate the best available information on marine ecosystems to foster knowledge, contribute to wise choices, and subsequently fulfill the needs and interests of a wide variety of sectors.

Methodology and General Description of North American Marine Ecoregions

In their coordinating role, the CEC and its partners brought together a trinational group of experts from throughout the continent to help define the Marine Ecoregions of North America—a nested series of ecoregions that provide a consensus framework for regional cooperative conservation efforts. Experts from an array of sectors—governmental agencies, NGOs, academic institutions and scientific research centers having expertise in a wide range of disciplines relating to marine science and planning—took part in the development process. This trinational, multi-sectoral development process included three stages: 1) individual country drafting efforts; 2) a trinational workshop to define the ecoregional classifications, agree on the criteria to delineate polygons for mapping the ecoregions, and 3) a peer review process. Engaging in a process that fostered discussion, debate and broad consensus among peers, the experts built their work on existing ecoregional descriptive frameworks (e.g., see the table on pages 21 ff., below) and used scientific data and information to support their decisions.

The following principles and general rules have guided development of the Marine Ecoregions of North America:

- The ecoregional map and its corresponding description are a trinational collaborative effort, tailored to particular needs and requirements, and based largely on expert knowledge and existing frameworks as well as the best available scientific data.

- The ecoregional map and its description includes three nested levels that link the global and more regional or local perspectives. The nested system reflects the nature of marine systems and the complex set of forces, pressures and threats that affect them, and will also allow North Americans to simultaneously address marine conservation problems of various geographic scales and differing scope at each level.

- The regional scale of the ecoregional description has been developed largely for and limited to North American waters within the Exclusive Economic Zone (EEZ). Levels I and II extend from the coastline to the outer edge of the EEZ. Level III covers an area from the coastline to the shelf edge or the 200-meter isobath on oceanic islands. While the EEZ was used to define the outer seaward limit of each country's territorial waters, it is acknowledged that ecosystems do not stop at political borders such as these.[3] The map focuses on maritime waters of continental North America, but also includes the US state of Hawaii as well as the territories of Puerto Rico and the US Virgin Islands. For completeness, a description of more distant territories has been included in the Appendix.

■ The three-dimensional nature of the ocean has been considered to the greatest degree possible in defining and delineating the different levels of ecoregions.

■ The map is not intended to specifically outline habitat type or substrate type, etc., but rather is intended to characterize ecosystems based on an alignment of selected characteristics at each level, distinguishing areas that may benefit from similar types of management and conservation measures. The framework thus allows for appropriate conservation strategies at the local, regional or continental level.

■ Most variables used to define the ecoregions are oceanographic or physiographic, reflecting the range of conditions that influence species distribution, and serve as practical surrogates for biological data that are largely incomplete or inconsistent in format at the North American scale. When available (such as at Levels I and III), information on faunal assemblages and community types was also used to help define the boundaries.

■ Each chapter is prefaced by a geographically limited extract from the map, showing that particular ecological region at Level I but also "zoomed down" to reveal the locations and approximate boundaries of Levels II and III. (See "About the map" for more detail on understanding these zoomed views.)

Within the Marine Ecoregions of North America, **Level I** captures ecosystem differences at the largest scale, defining large water masses and currents, large enclosed seas, and regions of coherent sea surface temperature or ice cover. This level is determined by processes that pertain to a whole ocean basin. The cross-shelf domain of Level I extends from the coasts to the deep oceans, although biogeographic patterns and processes in the deeper regions are still poorly understood. As a practical matter, the seaward boundary extends only to the jurisdictional limits of the Exclusive Economic Zone (EEZ), 200 nautical miles (370 km) offshore. The biogeographic regions themselves, however, may extend to beyond the EEZ.

It is important to note that the descriptions in each chapter of this book pertain to Level I. Detailed descriptions of Levels II and III will be found on the CEC website: http://www.cec.org.

Level II captures the break between neritic (near shore) and oceanic areas and is determined by large-scale physiography (continental shelf, slope, and abyssal plain, as well as areas of oceanic islands and major trenches, ridges and straits). This level reflects the importance of depth as a major determinant of benthic marine communities as well as the importance of major physiographic features in determining current flows and upwelling. Like Level I, it extends from the coasts to the EEZ.

Level III captures differences within the neritic realm and is based on more locally significant variables (local characteristics of the water mass, regional landforms, as well as biological community type). Level III is limited to the continental shelf, since only this area has sufficient information for finer-scale delineation.

About the map

The oceanic areas around North America have been mapped to display areas of general similarity in ecosystems and in the type, quality, and quantity of environmental resources. These mapped units are called marine ecological regions or ecoregions, which can serve as a spatial framework for research, assessment, management, and monitoring of ecosystems and the biota or other components of the ecosystems. There are three levels of spatial units provided in the ecoregions map (and the zoomed views shown for the discussion of each ecoregion)—the largest of which are the Level I ecoregions. Levels II and III are nested within the Level I ecoregions and are intended to provide successively greater descriptive precision: Level II represents major geomorphological features within the Level I ecoregions (e.g., the continental shelf, slope, etc.) and Level III represents the smallest-scale neritic units within the Level I shelf.

The identifiers for the ecological regions are a simple numerical shorthand: i.e., 19. 4 identifies the fourth Level II seafloor region within Level I ecoregion 19; 5.3.2 indicates the second Level III coastal region found within the third seafloor region (Level II) of ecoregion number five (Level I).

Different colors were assigned according to overall oceanic surface temperatures to help the reader visually distinguish between adjacent Level I ecological regions. Blues were used for the colder waters of the northern ecoregions, reddish hues for the warm, southern waters, and greenish or brownish hues for the temperate waters in between. Level II seafloor regions are differentiated by different color densities: generally they are darker for areas near shore and lighter for those farther from shore, depending on the prevailing water depth. Solid lines delineate Level III coastal regions from the Level II seafloor regions underneath.

---

3 In certain areas, associated adjacent waters outside the EEZ were included for the sake of completeness (e.g., the area around the Grand Banks).

The Marine Ecoregions map is based on the CEC's North American Altas framework that employs a Lambert Azimuthal Equal Area Projection centered on the North American continent. As a result, the locations and orientations of outlying insular areas (e.g., the Hawaiian Archipelago or the US Caribbean islands) may appear to be distorted when compared to other map projections. The outer boundaries shown on the map (and the zoomed views taken from it) are approximate and illustrative. They do not necessarily reflect the boundaries of the EEZ claimed by the three countries.

Limitations of the framework

There were many restrictions and constraints in developing the Marine Ecoregions of North America. Time and resources were limited, spatial and content information was often scarce or inconsistent.

The map and the zoomed views taken from it are simplified depictions of the distribution of marine ecosystems. For example, although maps typically show general spatial distribution of systems well, they do not commonly portray processes, structure, functions and continuous change very clearly. As well, map unit boundaries of ecosystems are shown as lines, whereas in nature they are best thought of as transition areas of quite variable width; in the marine realm, boundaries are more fluid, and processes are even more dynamic in time and space than they are on land. Furthermore, maps tend to show regions in two-dimensions—length and width. Mapped areas in the ocean, perhaps even more than typically on land, have to capture the third (vertical) dimension through descriptions. As a consequence, portraying marine ecosystems by a static, two-dimensional map was challenging. While the vertical extension of the classification extends from the *supra*-tidal zone of the coasts and wetlands (the splash, spray and aerosol zone) to the benthic bottom of the marine environment, Level I ecoregions generally reflect the biogeographic divisions of the continental shelf and the upper reaches of the pelagic zone, and borders have been extended to the edge of the EEZ rather arbitrarily. Major seas, such as the Gulf of Mexico and Bering Sea and major currents such as the Gulf Stream, each occupy single ecoregions. The fact that the surface biogeography does not cleanly convey information on habitat associations occurring in deeper areas is an issue that remains to be resolved.

The reader will note that there is some repetition in Level II region nomenclature. This is because some benthic features such as the Grand Banks (6.3, 7.1), the Mesoamerican Trench (16.3, 17.3, 18.6), and the East Pacific Rise (17.4, 18.5) were artificially sectioned due to their extent and location straddling different Level I ecological regions.

The CEC's classification of North American marine ecoregions was developed to facilitate trinational cooperation in marine conservation and was carried out after careful study of existing marine biogeographic systems and Large Marine Ecosystems and is built upon a rich literature of biogeographic marine classification. The following table compares the Level I marine ecoregions of North America with several of the most commonly used systems. The CEC's nested framework correlates well with systems defined by faunal distributions (e.g., Hayden et al. 1984), but represents regions and their distribution of biodiversity at a finer scale than do the Large Marine Ecosystems (LMEs, Sherman and Alexander 1986, Sherman and Duda 1999) or Biomes and Provinces (Longhurst 1998). In general, the classification system used herein corresponds and nests well with both coarser and finer-scale systems.

Finally, we should note that although this work focuses upon the marine ecoregions of North America, understanding the influences of landscapes and land use activities are vital as well. The work that the CEC undertook (CEC 1997) in describing the terrestrial ecoregions is a useful companion document to *Marine Ecoregions of North America*.

Relationship of Other Marine Frameworks and the Marine Ecoregions of North America

Comparable Marine Classifications

Marine Ecoregion Number	Marine Ecoregions of North America (this document)	Marine Ecoregions (Spalding et al. 2007)	Large Marine Ecosystems (LME) (Sherman and Duda 1999)	Oceanic and Coastal Realms and Provinces (Hayden et al. 1984)	Biogeographic Regions (US National Estuarine Research Reserve System, NOAA 1998)	Coastal Biogeographic Provinces and Regions (Sullivan Sealey and Bustamante 1999)	Ocean Biomes and Provinces (Longhurst 1998)	Conservation Planning Regions (WWF Canada in review[4])	National Marine Conservation Areas (Mercier and Mondor 1995[5])	Ecozones of Canada (Wiken et al. 1996)
1	Bering Sea	Arctic Realm: Eastern Bering Sea	Eastern Bering Sea LME	Arctic Province	Subarctic	—	Pacific Polar Biome; Bering Sea portion of the North Pacific Epicontinental Sea Province	—	—	—
2	Beaufort/Chukchi Seas	Arctic Realm: Beaufort Sea, eastern parts of Chukchi Sea, western portion of Beaufort-Admundsen-Viscount Melville-Queen Maud	Beaufort and Chukchi LMEs	Arctic Province	—	—	Atlantic Polar Biome; Boreal Polar Province	Mackenzie Delta and Beaufort Sea	Beaufort Sea	Western portion of Arctic Archipelago
3	Arctic Basin	Arctic Realm	Arctic Ocean LME	Arctic Province	—	—	Atlantic Polar Biome; Boreal Polar Province	Western Arctic, Gulf of Boothia and Foxe Basin	Arctic Basin	Arctic Basin
4	Central Arctic Archipelago	Arctic Realm: northern portion of Beaufort-Admundsen-Viscount Melville-Queen Maud, High Arctic Archipelago	Arctic Ocean LME	Arctic Province	—	—	Atlantic Polar Biome; Boreal Polar Province	Western Arctic, Gulf of Boothia and Foxe Basin	Arctic Archipelago	Northern portion of Arctic Archipelago
5	Hudson/Boothian Arctic	Arctic Realm: Lancaster Sound, Hudson Complex, southern portions of Beaufort-Admundsen-Viscount Melville-Queen Maud	Hudson Bay LME	Arctic Province	—	—	Atlantic Polar Biome; Boreal Polar Province	Mackenzie Delta and Beaufort Sea; Western Arctic, Gulf of Boothia and Foxe Basin; Central Canadian Arctic; Hudson and James Bays	Lancaster Sound, Queen Maud Gulf, Foxe Basin, Hudson Bay, James Bay	Southern portion of Arctic Archipelago
6	Baffin/Labradoran Arctic	Arctic Realm: Baffin Bay-Davis Strait, Northern Labrador, Northern Grand Banks-Southern Labrador	Newfoundland-Labrador Shelf LME	Arctic Province and part of the Acadian Province	—	—	Atlantic Polar Biome; Boreal Polar Province	Central Canadian Arctic; Ungava Bay, Hudson and Davis Straits; Newfoundland and Labrador Shelf	Lancaster Sound, Baffin Island Shelf, Hudson Strait, Labrador Shelf, Newfoundland Shelf, part of the Grand Banks	Northern portion of Northwest Atlantic
7	Acadian Atlantic	Temperate Northern Atlantic Realm; Cold Temperate Northwest Atlantic Province: Gulf of St. Lawrence-Eastern Scotian Shelf, Southern Grand Banks-South Newfoundland, Scotian Shelf, Gulf of Maine/Bay of Fundy	Southern portion of Newfoundland-Labrador Shelf LME; Scotian Shelf LME; northern portion of Northeast US Continental Shelf LME	Acadian Province	Acadian Region	—	Atlantic Coastal Biome; Northwest Atlantic Shelves Province	Part of the Northwest Atlantic; Bay of Fundy and Gulf of Maine; Scotian Shelf; Gulf of St. Lawrence, Grand Banks	Grand Banks, North Gulf Shelf, Laurentian Channel, St. Lawrence Estuary, Magdalen Shallows, Scotian Shelf, Bay of Fundy	Southern portion of Northwest Atlantic

4   As per The Nature Audit <http://assets.wwf.ca/downloads/thenatureaudit_may2003.pdf> as well as M. King, J. Smith and J. Laughren, pers. comm. WWF Canada's Conservation Planning Regions also correspond roughly to their *Biogeographic Subregions* for their work in their Northwest Atlantic Ecoregion.
5   Many of Parks Canada 29 National Marine Conservation Areas, defined by Mercier and Mondor 1995, closely approximate Level III Marine Ecoregions.

Marine Ecoregion Number	Marine Ecoregions of North America (this document)	Marine Ecoregions (Spalding *et al.* 2007)	Large Marine Ecosystems (LME) (Sherman and Duda 1999)	Oceanic and Coastal Realms and Provinces (Hayden *et al.* 1984)	Biogeographic Regions (US National Estuarine Research Reserve System, NOAA 1998)	Coastal Biogeographic Provinces and Regions (Sullivan Sealey and Bustamante 1999)	Ocean Biomes and Provinces (Longhurst 1998)	Conservation Planning Regions (WWF Canada in review[4])	National Marine Conservation Areas (Mercier and Mondor 1995[5])	Ecozones of Canada (Wiken *et al.* 1996)
8	Virginian Atlantic	*Temperate Northern Atlantic Realm: Cold Temperate Northwest Atlantic Province;* Virginian	Northeast US Continental Shelf LME	Virginian Province	Virginian Region	—	Atlantic Coastal Biome; Northwest Atlantic Shelves Province	—	—	—
9	Northern Gulf Stream Transition	*Temperate Northern Atlantic Realm:* offshore portion of *Cold Temperate Northwest Atlantic Province*	Northeast US Continental Shelf LME	Virginian and Acadian Provinces	—	—	Atlantic Westerly Winds Biome; Gulf Stream Province	Bay of Fundy and Gulf of Maine; Scotian Shelf; Grand Banks	Scotian Shelf, Grand Banks	Southern portion of Atlantic (Ecozone)
10	Gulf Stream	*Temperate Northern Atlantic Realm: Warm Temperate Northwest Atlantic Province:* Carolinian	Southeast US Continental Shelf LME	Carolinean Province	—	—	Atlantic Westerly Winds Biome; North Atlantic Subtropical Gyral Province	—	—	—
11	Carolinian Atlantic	*Temperate Northern Atlantic Realm; Warm Temperate Northwest Atlantic Province:* Carolinian	Southeast US Continental Shelf LME	Carolinian Province	Carolinian Region	—	Atlantic Coastal Biome; Northwest Atlantic Shelves Province	—	—	—
12	South Florida/ Bahamian Atlantic	*Tropical Atlantic Realm; Tropical Northwestern Atlantic Province:* Floridian	Southeast US Continental Shelf LME; Gulf of Mexico LME; Caribbean Sea LME	Carolinian Province	West Indian Region	Tropical Northwestern Atlantic Biogeographic Province; South Florida & Bahamian Coastal Biogeographic Regions	Atlantic Trade Winds Biome; Caribbean Province	—	—	—
13	Northern Gulf of Mexico	*Temperate Northern Atlantic Realm; Warm Temperate Northwest Atlantic Province:* Northern Gulf of Mexico	Gulf of Mexico LME	Louisianian Province	Louisianian Region	—	Atlantic Trade Winds Biome; Caribbean Province	—	—	—
14	Southern Gulf of Mexico	*Tropical Atlantic Realm; Tropical Northwestern Atlantic Province:* Southern Gulf of Mexico	Gulf of Mexico LME	Caribbean Province	—	Tropical Northwestern Atlantic Biogeographic Province; Gulf of Mexico Coastal Biogeographic Region	Atlantic Trade Winds Biome; Caribbean Province	—	—	—
15	Caribbean Sea	*Tropical Atlantic Realm; Tropical Northwestern Atlantic Province:* Eastern Caribbean, Western Caribbean, Greater Antilles	Caribbean Sea LME	Caribbean Province	West Indian Region	Tropical Northwestern Atlantic Biogeographic Province; Central Caribbean Coastal Biogeographic Region	Atlantic Trade Winds Biome; Caribbean Province	—	—	—
16	Middle American Pacific	*Tropical Eastern Pacific Realm; Tropical East Pacific Province:* Chiapas- Nicaragua	Central portion of Pacific Central American LME	Panamanian Province	—	Tropical Eastern Pacific Biogeographic Province; Chiapas- Nicaragua Coastal Biogeographic Region	Pacific Coastal Biome; Central American Coastal Province	—	—	—

		Comparable Marine Classifications								
Marine Ecoregion Number	Marine Ecoregions of North America (this document)	Marine Ecoregions (Spalding et al. 2007)	Large Marine Ecosystems (LME) (Sherman and Duda 1999)	Oceanic and Coastal Realms and Provinces (Hayden et al. 1984)	Biogeographic Regions (US National Estuarine Research Reserve System, NOAA 1998)	Coastal Biogeographic Provinces and Regions (Sullivan Sealey and Bustamante 1999)	Ocean Biomes and Provinces (Longhurst 1998)	Conservation Planning Regions (WWF Canada in review[4])	National Marine Conservation Areas (Mercier and Mondor 1995[5])	Ecozones of Canada (Wiken et al. 1996)
17	Mexican Pacific Transition	*Tropical Eastern Pacific Realm; Tropical East Pacific Province:* Mexican Tropical Pacific, Clipperton, Revillagigedo	Northern portion of Pacific Central American LME	Mexican Province	—	Tropical Eastern Pacific Biogeographic Province; Mexican Tropical Pacific and Clipperton & Revillagigedo Islands Coastal Biogeographic Regions	Pacific Coastal Biome; Central American Coastal Province/ Pacific Trade Winds Biome; North Pacific Ecuatorial Countercurrent, North Pacific Tropical Gyre Provinces	—	—	—
18	Gulf of California	*Temperate North Pacific Realm; Warm Temperate Northeast Pacific Province:* Cortezian	Gulf of California LME; northeasternmost portion Pacific Central American LME	Cortezian and Mexican Provinces	—	Warm Temperate Northeastern Pacific Biogeographic Province; Cortezian Coastal Biogeographic Region	Pacific Coastal Biome; Central American Coastal Province	—	—	—
19	Southern Californian Pacific	*Temperate North Pacific Realm; Warm Temperate Northeast Pacific Province:* Southern California Bight, Magdalena Transition	Northwesternmost portion of the Pacific Central American LME; southern portion of the California Current LME	San Diegan Region and Mexican Province	Californian (Southern California subregion)	Warm Temperate Northeastern Pacific and Tropical Eastern Pacific Biogeographic Provinces; Mexican Tropical Pacific, Mexican Temperate Pacific, and Magdalena Transition Coastal Biogeographic Regions	Pacific Coastal Biome; southern portion of the California Current Province	—	—	—
20	Montereyan Pacific Transition	*Temperate North Pacific Realm; Cold Temperate Northeast Pacific Province:* Northern California	Central portion of the California Current LME	Oregonian Province	Californian Region (Central California & San Francisco Bay subregions)	—	Pacific Coastal Biome; central portion of the California Current Province	—	—	Southern portion of Pacific Marine
21	Columbian Pacific	*Temperate North Pacific Realm; Cold Temperate Northeast Pacific Province:* Oregon-Washington-Vancouver Coast and Shelf, Puget Trough/Georgia Basin	Northern portion of the California Current LME	Oregonian Province	Columbian Region	—	Pacific Coastal Biome; northern extent of the California Current Province	Southern Queen Charlotte Sound and Strait of Georgia; United States Pacific Northwest Waters	Vancouver Island Shelf; Strait of Georgia	Northern portion of Pacific Marine
22	Alaskan/Fjordland Pacific	*Temperate North Pacific Realm; Cold Temperate Northeast Pacific Province:* Gulf of Alaska, North American Pacific Fjordland, Pacific Oceanic portion of Aleutian Islands	Gulf of Alaska LME; Pacific portion of Eastern Bering Sea LME	Aleutian and Sitkan provinces & northern portion of Oregonian Province	Fjord Region (Southern Alaska subregion)	—	Pacific Coastal Biome; Alaska Downwelling Coastal Province	Gulf of Alaska; Northern Queen Charlotte Sound and Southeastern Alaskan Waters; Southern Queen Charlotte Sound and Strait of Georgia	Queen Charlotte Sound; Queen Charlotte Shelf; Hecate Strait	—
23	Aleutian Archipelago	*Temperate North Pacific Realm; Cold Temperate Northeast Pacific Province:* Aleutian Islands	Eastern Bering Sea LME	Aleutian Province	Fjord Region (Aleutian Islands subregion)	—	Pacific Coastal Biome; Alaska Downwelling Coastal Province	—	—	—
24	Hawaiian Archipelago	*Eastern Indo-Pacific Realm;* Hawaii Province	Insular Pacific -Hawaii LME	—	Insular Region (Hawaiian Islands)		Pacific Trade Winds Biome; North Pacific Tropical Gyre Province/North Pacific Transition Zone Province	—	—	—

Schools of prized coho salmon rely on Pacific waters for much of their lives but are dependent upon inland freshwaters to spawn.
Photo: Brandon D. Cole

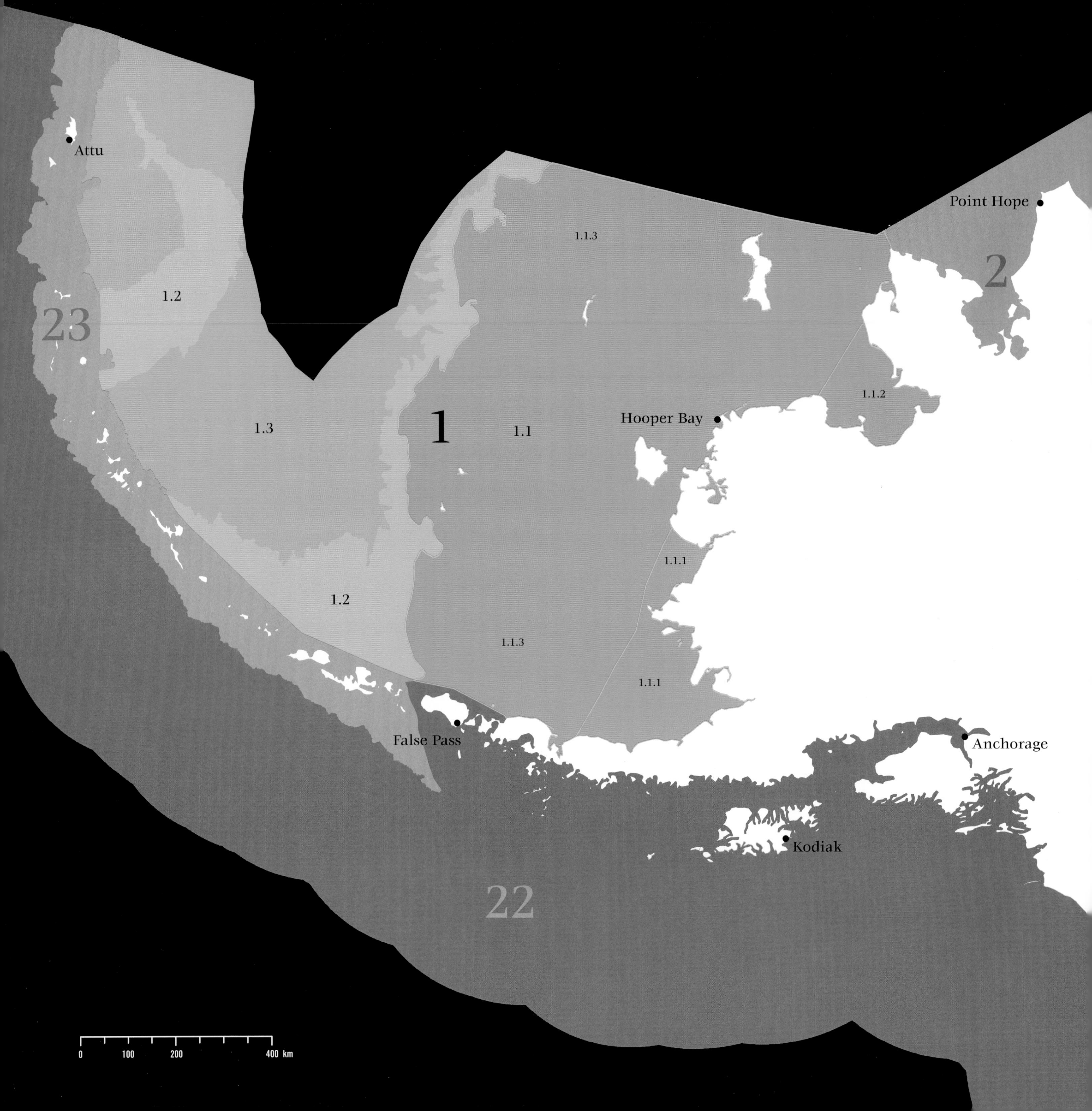

23

Attu

1.2

1.3

1.2

1

1.1

1.1.3

Hooper Bay

1.1.2

1.1.1

1.1.3

1.1.1

Point Hope

2

False Pass

Kodiak

Anchorage

22

0   100   200   400 km

1. Bering Sea

Level II seafloor geomorphological regions include:

    1.1   Bering Shelf
    1.2   Bering Slope and Bowers Ridge
    1.3   Bering Basin

Level III coastal regions include:

    1.1.1   Bristol Bay and Kuskokwim Bay
    1.1.2   Norton Sound
    1.1.3   Middle and Outer Bering Sea Neritic

Regional Overview

The Bering Sea is the world's third-largest, semi-enclosed body of water. Noted in particular for its wide coastal shelf and high productivity, the Bering Sea Ecoregion is of special conservation importance to marine mammals and fisheries, and is a unique subpolar ecosystem. It is bounded by the Bering Strait in the north and the arc of the Aleutian Island chain to the south, and divided in half physiographically, with a broad shelf to the east and much deeper oceanic plains to the west.

The Bering Sea Region includes two B2B[6] Marine Priority Conservation Areas (PCAs): PCA 1-Pribilof Islands, and PCA 2-Bristol Bay (Morgan *et al*. 2005).

Physical and Oceanographic Setting

The Bering Sea Ecoregion includes a broad eastern shelf, gently sloping for the first 100 m and then descending at a slightly steeper rate to the shelf edge at around 200 m. The continental slope is incised with many canyons and drops to a generally flat abyssal plain, 3,700–4,000 m deep. Major islands on the shelf are the St. Lawrence and Nunivak Islands (which are the largest in the Bering Sea), the Pribilofs (Fur Seal Islands), and St. Matthew, Nelson, and Karagin Islands. Currents in the Eastern Bering Sea flow in a generally counter-clockwise direction, with an Aleutian North Slope Current flowing northeast along the inside of the Aleutian Island chain and curving northwest at the shelf edge to form the Bering Slope Current. A northeast drift current flows along the coast over the nearshore shelf. Net circulation flow is from the Bering Sea to the Chukchi Sea through the Bering Strait. The annual formation and retreat of sea ice through the Bering Strait and out over the northeast shelf is a major determinant of the distribution of many species. The Yukon River is the largest river flowing into the eastern Bering Sea.

Biological Setting

The Bering Sea is among the most productive of high-latitude seas, supporting a large biomass of fishes, birds and marine mammals. The southeast Bering Sea contains two fairly distinct upper-trophic level species groups, or guilds, based on characteristics of feeding. The first group consists of an outer shelf pelagic group of

---

6   As noted by Morgan and others (2005), "a total of 28 sites were identified as PCAs, spanning seven marine ecoregions in the B2B region, totaling eight percent of the total EEZ area of the three nations." PCAs for the Arctic and Atlantic Oceans and the Caribbean Sea have not as yet been identified. For further information on the process and the regions identified, see Morgan, L., S. Maxwell, F. Tsao, T.A.C. Wilkinson and P. Etnoyer. 2005. Marine Priority Conservation Areas: Baja California to the Bering Sea. Montreal: CEC and MCBI.

Fact Sheet

Rationale: defined by sea surface temperature and the semi-enclosed physiography of the sea

Surface: 1,468,220 km²

Sea surface temperature: Avg. <2°C (winter), 6°–14°C (summer)

Major currents and gyres: dominated by tidal flows. A counter-clockwise Aleutian North Slope Current and Bering Slope Current flow along the north edge of the Aleutians and west edge of the Bering Shelf respectively. The major direction of flow is northward through the Bering Strait.

Physiography: The wide coastal shelf is bounded by the Aleutian Island chain to the south and the Bering Strait to the north.

Depths: shelf (roughly 0–200 m): 51%[7]; slope (roughly 200–2,500/3,000 m): 14%; abyssal plain (roughly 3,000+ m): 35%

Substrate type: generally muddy sand and gravel

Major community types and subtypes: seasonal sea ice, high productivity pelagic systems, polar and subpolar communities

Productivity: moderately high (150–300 g C/m²/yr)

Species at risk: bowhead whale, blue whale, fin whale, gray whale, North Pacific right whale, Pacific walrus, Steller sea lion, northern fur seal, short-tailed albatross, red-legged kittiwake, Steller's eider, king eider

Human activities and impacts: commercial and subsistence fishing and hunting, oil exploration and recovery

---

7   Percentages found in the fact sheet and following text are approximate. There are some cases where totals amount to 99 or 101 percent. This is due to the addition of rounded (up or down) percentages.

fish, mammals and birds that consume small fish—primarily juvenile pollock, and krill. The second group is an inshore group of fish, crabs and other bottom dwelling fauna that consume mainly benthic infauna. In the outer shelf, walleye pollock dominate the biomass and represent a keystone species in the system. Springer (1996) determined that juvenile walleye pollock are important prey for seabirds during the nesting season. In addition, Lowry *et al.* (1996) found that all species of phocid seals feed on walleye pollock, with the species predominating the diets of harbor, spotted and ribbon seals.

Many species move seasonally with the advance and retreat of sea ice, including many fishes, walrus and seals. Summer residents consist of species that feed and reproduce in the region, such as northern fur seal, Steller sea lion, and seabirds such as murres, black-legged kittiwakes, and auklets, as well as species that use the area to feed but reproduce elsewhere—including fin whale, the eastern North Pacific stock of gray whale and oceanic seabirds, especially shearwaters.

The region is particularly important for marine mammals—especially the rare Steller sea lion. The western Alaskan stocks

Crested and least auklets nest on island coasts such as those of the Aleutian Islands in the Bering Sea.   © Nikolay Konyikhov/AccentAlaska.com

of Steller sea lion have undergone a continuous decline since the 1970s—they were listed as a threatened species under the US Endangered Species Act in 1990, then declared endangered in 1997. Management measures, including buffer zones around rookeries and alternatives to mitigate potential impacts of commercial fisheries on important prey species, are presently being developed.

Over the last decade, significant changes have occurred in the Bering Sea, possibly due to shifts in the Pacific Decadal Oscillation and Arctic Oscillation. The character of the seasonal ice pack recently has changed from the "warm" phase that persisted since the regime shift of the late 1970s to one that exhibits rapid buildup in winter but earlier retreat in spring. The timing of spring primary production is determined predominately by the timing of ice retreat. The distribution and abundance of pollock, Pacific cod and other commercially important fishes have varied with these fluctuations in sea ice, and the ability of these fish populations to withstand fishing pressure is expected to vary between warm and cold regimes (Hunt *et al.* 2002).

Human Activities and Impacts

Several small towns and numerous Alaskan native villages sparsely populate the Alaskan coast of the Bering Sea. Subsistence use of coastal resources in the region includes nearshore harvesting of salmon and other fishes, invertebrates and marine mammals, including seals, Steller sea lion, Pacific walrus and polar bear. Subsistence hunting of the northern fur seal on the Pribilof Islands is limited by quota and other regulations as the species has been listed as depleted in the US Marine Mammal Protection Act. Offshore, commercial fisheries in the region are of major importance—Bering Sea fish and shellfish constitute almost 5 percent of the world and 40 percent of the US fisheries harvest (Macklin 1999). Pollock, salmon, Pacific halibut and crab generate over US$2 billion each year in fisheries revenue and provide a major source of protein. At present, some Bering Sea fisheries, such as pollock, have experienced major changes in abundance over the last thirty years, although they appear not to be overexploited. Populations of several species, such as king crab and Greenland halibut, however, are at near historical lows. In addition to fishing and hunting, oil and gas exploration and recovery activities are prevalent in the region. Ecological conditions of the coastal resources in Alaska are poorly known. Alaska has assessed less than 0.1 percent of its coastal estuaries (EPA 2005). In the US northern Pacific and western arctic ecoregions (1, 2, 22 and 23) two out of 35 federally fished stocks are overfished (NMFS 2007).

Adult red-legged kittiwake at nest. One of the largest colonies of this species, currently at risk, is in the Pribilof Islands.
© Nikolay Konyikhov/AccentAlaska.com

Male Pacific walrus haul out on the rocky shores of Round Island, Alaska, to rest and sunbathe. *Photo:* Tom Bledsoe/DRK PHOTO

1

2

2.1.3

2.1

2.1.1

Point
Hope

2.1

2.1.4

Prudhoe Bay

3

4

2.1.2

2.1

2.1.4

2.1.4

2.2

Holman

5

Anchorage

22

0   50 100   200 km

2. Beaufort/Chukchi Seas

Level II seafloor geomorphological regions include:

2.1   Beaufort/Chukchian Shelf
2.2   Beaufortian Slope

Level III coastal regions include:

2.1.1   Kotzebue Sound
2.1.2   Mackenzie Estuarine Area
2.1.3   Chukchian Neritic
2.1.4   Beaufortian Neritic

Regional Overview

This sparsely populated area, particularly well known for its coastal oil and gas activities, is also home to 40 species of fish and significant concentrations of marine mammals like the bowhead whale, polar bear and ringed seal. The Beaufort/Chukchi Ecoregion borders the Arctic Ocean and is shared by Canada, Russia and the United States. It is bounded by the Bering Strait in the southwest, permanent sea ice of the Arctic Basin Ecoregion, and follows the coastal shelf along the north shore of Alaska and Canada's Yukon and Northwest Territories to Amundsen Gulf.

Physical and Oceanographic Setting

The Beaufort/Chukchi Seas Ecoregion includes waters of the shallow Chukchi Sea (depths of 0–100 m) and Beaufort Sea, which has a narrow, shallow shelf along Alaska, and a broad (extending 100 km offshore) yet still shallow shelf off the Yukon and Northwest Territories—where depths of 10 m or less are found up to 30 km from shore. The deepest point in the eastern part of this ecoregion is the Amundsen Gulf (600 m in the center), characterized by large bays and little shallow water. The ecoregion's continental slope drops steeply to the Arctic Basin. The more coastal areas of the ecoregion are composed of a series of barrier beaches, spits, extensive deltas, lagoons, estuaries, tidal flats and narrow sand and gravel beaches, with a low coastal relief and a generally wide shelf. Minimal tidal influence also characterizes the ecoregion.

Circulation is driven by Bering Sea water flowing northward through the Bering Strait, which strongly influences the southern Chukchi Sea and contributes to a summer coastal ice-free zone 150–200 km wide. The ecoregion is covered with a combination of landfast ice, stretching 20–80 km from the shore, and the rest by pack ice from October to June—a factor that contributes to low salinities, as well as most of the biological characteristics of the ecoregion. Two additional important physical features worth noting are the Cape Bathurst Polynya and a smaller polynya in Lambert Channel that appear in the spring. The Mackenzie River Delta also influences this ecoregion, particularly its bottom sediment characteristics (sandy to silty), low salinity and high water turbidity. Currents of the ecoregion are moderate to strong, while tidal range is small and tidal flows are weak. Sea surface temperatures in summer are below 12°C and average 8°C in the southwest and along the Beaufort coast and colder to the north.

Fact Sheet

Rationale: defined by sea surface temperature and a transition between boreal and Arctic faunas

Surface: 446,009 km²

Sea surface temperature: <12°C (summer), 8°C (average) in the southwest and along the Beaufort coast

Major currents and gyres: Cape Bathurst Polynya. A smaller polynya in Lambert Channel appears in the spring.

Depth: shelf (roughly 0–200 m): 88%; slope (roughly 200–2,500/3,000 m): 12%; abyssal plain (roughly 3,000+ m): 0% Note that the deepest point in the Canadian area of this region is the Amundsen Gulf (600 m in the center).

Substrate type: sandy to silty, sand and gravel beaches

Major community types and subtypes: seasonal sea ice, polar and subpolar communities, coastal wetlands and delta communities

Productivity: moderately high (150–300 g C/m²/yr), but only in the summer when the ice melts[8]

Species at risk: polar bear, bowhead whale, beluga whale, and gray whale

Human activities and impacts: oil and gas, commercial fisheries, mining

8 On the basis of bathymetry and productivity, this region has been partitioned into a separate low productivity Beaufort Large Marine Ecosystems (LME) and shallow, moderately high productivity Chukchi LME. The LME delineation of this heterogeneous region may therefore be more useful for certain purposes.

Biological Setting

In Arctic terms, this ecoregion can be considered of moderately high biological productivity due mostly to the mixing of the freshwaters from the Mackenzie River with the salt waters of the Beaufort Sea. However, its productivity is lower than many other portions of the Pacific or Atlantic Oceans. Sea ice is the most important seasonal feature affecting the ecoregion's fauna. Retreating sea ice in early summer contributes to plankton blooms that provide the major productivity for the ecosystem. Most marine mammals are ice-associated, following the advance and retreat of the sea ice from the Beaufort and Chukchi Seas through the Bering Strait and into the Bering Sea.

The ecoregion has relatively high fisheries production and provides habitat to 40 species of fish, including capelin, Pacific herring, Greenland cod, Arctic cod, species of the genus *Coregonus* such as cisco and whitefish; and sea urchins, mussels and mollusks, sea cucumbers, sea stars and anemones. It also has significant concentrations of marine mammals, such as bowhead and gray whales, polar bear, and ringed and bearded seals. The Beaufort/Chukchi Ecoregion includes the northern extent of some north Pacific boreal fauna—for example, salmon, Pacific herring, walleye pollock—especially in the Chukchi Sea. The Mackenzie Delta provides a very important wetland and migratory bird habitat for species such as king eider, long-tailed duck, scoters, mergansers, scaup, loons, geese (Chen, Branta and Anser species) and shorebirds such as red-necked phalarope. The only breeding populations of thick-billed murre and black guillemot in the western Arctic are found in this region. Overall, however, the ecoregion has significantly lower densities of fishes and seabirds than the Bering Sea Ecoregion.

Human Activities and Impacts

The ecoregion is well known for its coastal oil and gas activities, particularly on Alaska's North Slope along the Alaskan Arctic Coast, which provides a major source of employment and income for the state. The region's harsh environment makes it unsuitable for most other economic activities. Although there have not been major reports of damage to marine ecosystems from activities associated with the oil and gas exploitation, new and renewed interests in exploration—within Arctic National Wildlife Refuge and offshore in the Canadian Beaufort Sea, for example—are believed by many to be a cause for concern.

The coast is sparsely populated, predominantly by small communities of Alaskan native peoples, Inuvialuit and other indigenous groups. Subsistence use of coastal resources is particularly important to these communities and includes nearshore harvesting of salmon and other fishes, invertebrates and marine mammals (including seals and Steller sea lion, walrus, bowhead whale and polar bear). In the US northern Pacific and western arctic ecoregions (1, 2, 22 and 23) two out of 35 federally fished stocks are overfished (NMFS 2007).

The thick-billed murre, one of the most common seabirds in Canada's North, nests in large colonies on sheer cliffs too hazardous for terrestrial predators.   © Nikolay Konyikhov/AccentAlaska.com

The bowhead whale is one of the two species of right whales and is commonly found in shallower waters and near pack ice.   *Photo:* Flip Nicklin/Minden Pictures

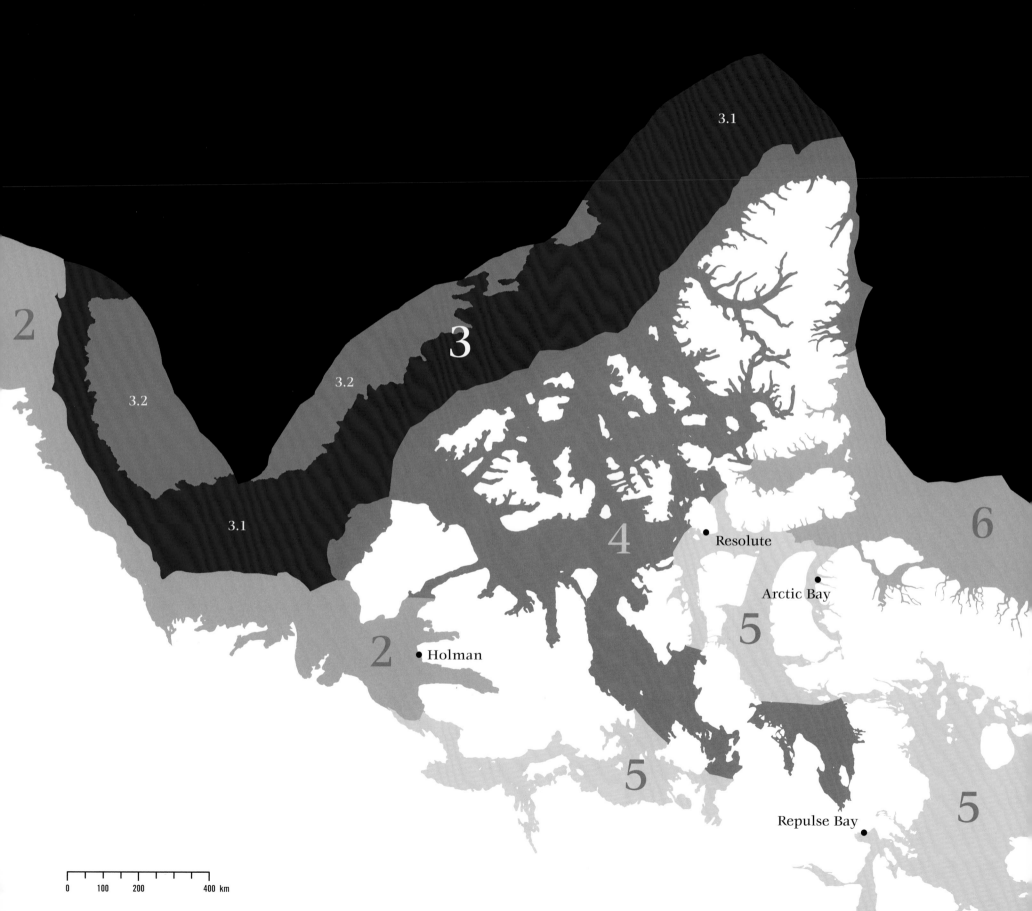

2

3

3.1

3.2

3.2

3.1

4

5

5

6

5

• Resolute

• Arctic Bay

2 • Holman

• Repulse Bay

0   100   200   400 km

3. Arctic Basin

Level II seafloor geomorphological regions include:

3.1   Arctic Slope
3.2   Arctic Plains

No Level III coastal regions are found in this ecoregion.

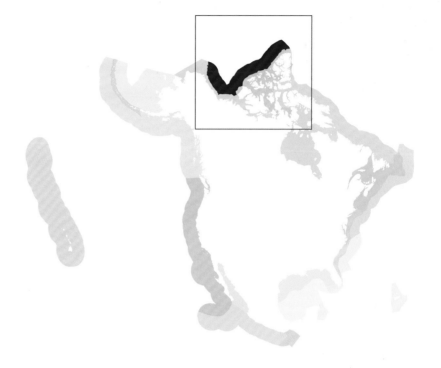

Regional Overview

The Arctic Basin Ecoregion is essentially the core northern part of the Arctic Ocean that remains under permanent ice cover—on the surface; it resembles land more so than ocean. Blooms of phytoplankton occur in spring and summer along the edges of the pack ice and algae grow on the underside of sea ice. These blooms are the basis of some of the Arctic food chains. The ecoregion encompasses a large, deep depression that reaches 3,600 m in depth, with no coasts. Like much of the Arctic, the Arctic Basin remains relatively isolated, poorly explored and under described.

Physical and Oceanographic Setting

The distinctive characteristic of this ecoregion is the relatively constant cover of ice sheets and pack ice. This is an expression of the very cold seawater temperatures, the northerly latitudes and the restricted influence of warmer southern waters. This giant permanent ice cap floats on the Arctic Ocean and covers more than 90 percent of the ecoregion and its extension into the Canada Basin. Driven by the Arctic Ocean Gyre, it slowly rotates counter-clockwise, roughly centered on the North Pole. In this ecoregion, ice can become heavily ridged and greater than 2 m thick. Islands of ice several kilometers square are also common.

The water column is relatively stable, with a permanent layer of relatively low salinity in the upper 100 m. This sets up a strong salinity-based vertical structure that restricts primary productivity during the summer period due to nutrient limitations. This stratification, too, establishes horizontal density gradients that are responsible for the surface ocean currents. Because of very large submarine ridges, the deep basin waters are walled in, largely isolated from adjacent waters. As a result, a stagnant pool of very cold water (roughly -1°C) lies at the bottom of the Arctic Basin.

The undersea geography is dominated by the Canada Basin, which plummets to an average depth of about 3,600 m. The Canada Basin extends from the Beaufort Sea almost to the North Pole, where it is bounded to the north by the Alpha-Mendeleev Ridge, the Lomonosov Ridge, and Nansen-Gakkel Ridge—submarine mountain ranges below the ocean surface.

The climate is extremely cold and dry. In January, mean daily temperatures range from -30° to -35°C. In summer, the mean daily temperature rises only to about 5°C. Annual precipitation ranges from 100 to 200 mm. Winds are generally out of the west or northwest, although there are local and seasonal variations. For instance,

Polar bears roam and hunt across a vast mosaic of frozen and open arctic waters.   *Photo:* Patricio Robles Gil

Fact Sheet

Rationale: ice regimes (faunal assemblages as a result)

Surface: 911,771 km²

Sea surface temperature: largely permanent ice in the long winter as well as short summer seasons

Major currents and gyres: the Arctic Ocean Gyre/the Beaufort Gyre

Other oceanographic features: ice covers 90–100 percent of the ecoregion in any given year. Ice cover over the year is not continuous, however, and numerous leads of open water do occur.

Depth: shelf (roughly 0–200 m): 0%; slope (roughly 200–2,500/3,000 m): 73%; abyssal plain (roughly 3,000+m): 27%. Note that to the west of the Queen Elizabeth Islands, the Canada Basin plummets to an average depth of about 3,600 m, and adjacent to the North Pole, it rises to 1,000 m depth at the Lomonosov Ridge, a narrow, submarine mountain range.

Major community types and subtypes: ice algae and phytoplankton are important primary producers; Arctic cod, Arctic sculpins, Arctic eelpouts and snailfish are present; whales are rare, and polar bear and ringed seal are the main marine mammals; Arctic benthic organisms such as anemones, clams, sea worms, sea stars and sponges are also present

Productivity: Moderate to low

Species at risk: polar bear

Key habitat: polynyas provide important feeding grounds for marine mammals and birds and serve as "islands" of high productivity within a sea of ice

Human activities and impacts: pesticides used in agriculture in southern and western latitudes are carried by wind to northern latitudes, including the Arctic Ocean (global distillation effect)

in the western portion of this ecoregion, winds are more evenly distributed from the southeast and northwest.

Biological Setting

The Arctic Basin Ecoregion is characterized by low productivity, in comparison to other, more southerly and warmer marine systems. This region's productivity is limited primarily by the availability of sunlight as well as by low year-round cold water temperatures and the very brief summer period (Wiken *et al.* 1996). However, phytoplankton and algae survive in this cold setting and have adapted to life on or near the permanent ice pack. Biological hotspots consisting of blooms of phytoplankton occur in spring and summer along the edges of the pack ice and algae grows on the underside of sea ice. These blooms provide the foundation for many of the Arctic food chains.

Although not typical throughout the area, several species live along the southerly margins of this region. They include walrus, polar bear, beluga whale, narwhal, and bearded, harp, harbor and ringed seals, the last being the main prey species of the polar bear. Migratory birds pass through the ecosystem, and a hardy Arctic species, the ivory gull, migrates to Davis Strait in the winter. This bird species will eat anything it can find, including fish, crustaceans, and even whale and seal carcasses.

Life is present beneath the ice too, but most species are less densely distributed than in the ice-free waters of the Pacific and Atlantic oceans. About 130 species of fish occur across the Arctic. The greatest numbers occur in the west and south, with schools of Arctic cod and Greenland cod, Arctic char, Arctic sculpins, Arctic eelpouts, and snailfish the most common. It is estimated that half the living creatures in the Arctic are benthic organisms such as anemones, clams, sponges, sea worms, and sea stars. However, little research has been carried out in this region and little is known about these creatures other than that they are crucial to the Arctic food web. One of the most important of these species is the Arctic cod, which plays a key role in the Arctic marine food chain. It relies on planktonic organisms in the upper layers of the water column as opposed to the benthic organisms preferred by its close relative, the Atlantic cod. It is believed that Arctic cod is the basic diet of at least 12 species of marine mammals, 20 species of seabirds, and four species of fish.

Human Activities and Impacts

Much of the Arctic remains poorly explored and described, and the Arctic Basin is no exception. Some subsistence hunting and fishing exist, as well as limited oil and gas exploration. Aircraft, skidoos and ice-breaking ships have carried scientists and even tourists into parts of the area. Scientific expeditions have concentrated largely on finding oil and gas reserves along the edges of the ice pack—the permanent ice, however, presents great challenges to petroleum exploration and drilling.

In spite of its northerly isolation, urban and industrial areas and human activities far to the south affect this ecosystem and adjacent ecosystems—much more than is generally thought. Substances such as PCBs, DDT, and mercury, used and released in distant locations, are carried aloft by atmospheric circulation to the Arctic where they condense (global distillation effect), or they may arrive via ocean currents. These chemicals build up in the bodies of the marine mammals and people. In addition, global warming greatly threatens the ecology of the area, as ice packs and cover diminish in size and oceanic circulation patterns shift.

Arctic and polar cod thrive under the ice sheets covering the ocean in the higher latitudes of the Arctic. *Photo:* Elisabeth Calvert, UAF/NOAA

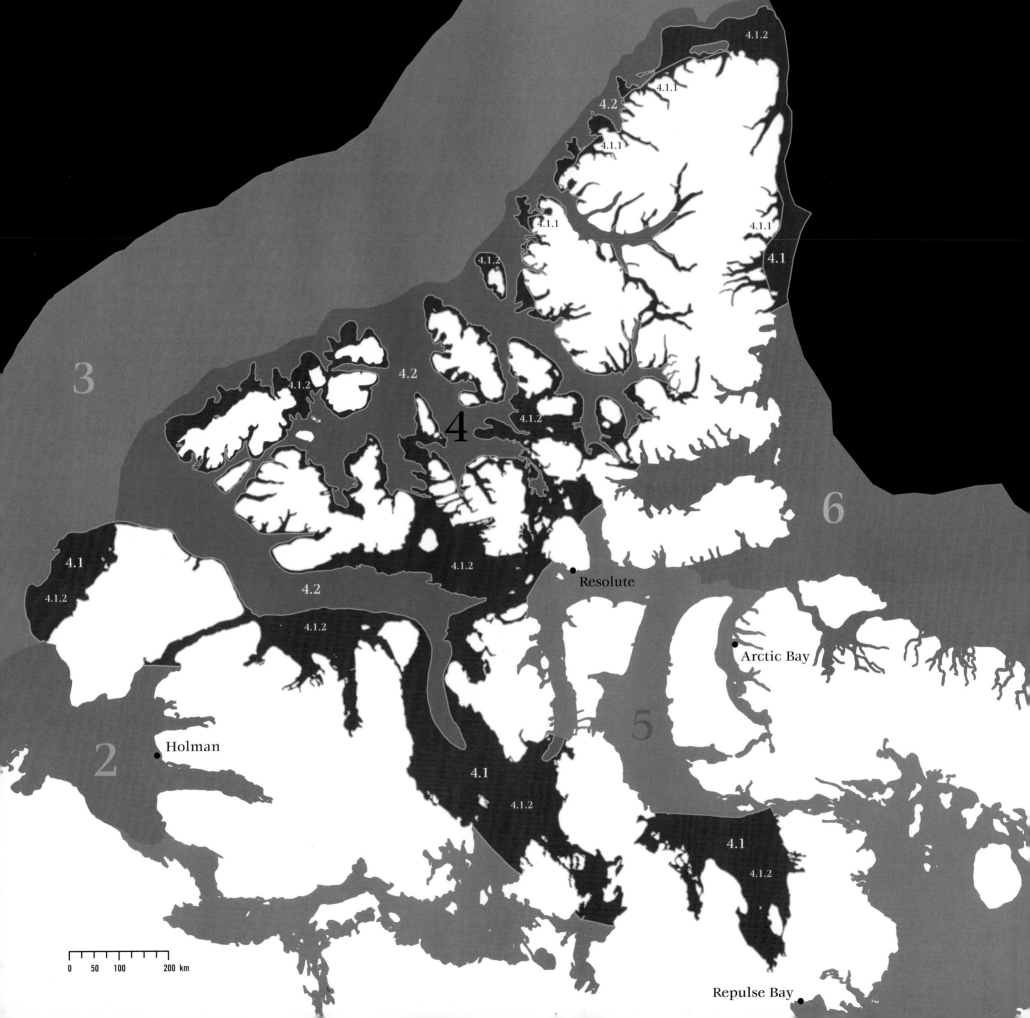

4. Central Arctic Archipelago

Level II seafloor geomorphological regions include:

4.1   Central Arctic Shelf
4.2   Central Arctic Slope

Level III coastal regions include:

4.1.1   Central Arctic Estuarine Areas
4.1.2   Central Arctic Neritic

Regional Overview

The Central Arctic Archipelago includes thousands of islands with jagged coastlines, making it one of the biggest archipelagos in the world and one of the areas containing the longest coastlines. The ecoregion is composed of waters mostly 200–500 m deep and includes most of the Arctic Islands east of the Arctic Basin Ecoregion, such as Ellesmere Island, the Queen Elizabeth Islands (with glacially scoured fjords 600–920 m deep) and the northeastern part of Victoria Island. During the brief Arctic summer, many of species of migrating birds are dependent on the ecoregion, making use of the unpredictable sections of open water that appear.

Physical and Oceanographic Setting

A patchwork of interconnecting bays, fjords, channels, straits, sounds and gulfs comprise the ecoregion, with very little of it composed of shallow waters. This lattice of marine water bodies surrounds the hundreds of islands that form the Queen Elizabeth chain. As with the neighboring Arctic Basin, this ecoregion's very cold seawater and northern latitude, as well as the little influence warmer southern waters have on the realm, make for its relatively constant cover of ice sheets and ice pack.

The water column is relatively stable, with a permanent layer of relatively low salinity in the upper 100 m. This sets up a strong salinity-based vertical structure that restricts primary productivity during the summer period due to nutrient limitations. This stratification also establishes horizontal density gradients that are responsible for the surface ocean currents. Tides in this ecoregion are minimal, as are the associated tidal currents.

During the winter, sea ice is jammed fast to the land and extends over the seas as a solid sheet. Polynyas, localized breaches in the ice where currents and upwellings create open water, can occur throughout the ecoregion. The ice cover reaches its maximum thickness in May, and in the brief spring and summer periods, the ice will break up. In the northwestern parts of the ecoregion, the sea ice normally shatters into massive sheets that are separated by narrow channels of open water. The sea ice persists throughout the summer. Along the east coast of Ellesmere Island, icebergs occasionally calve (break off) from adjacent coastal glaciers, making their way to the ocean.

As with the neighboring Arctic Basin Ecoregion, the climate of the Central Arctic Archipelago is extremely cold and dry.

Fact Sheet

Rationale: ice regimes (faunal assemblages as a result)

Surface: 673,054 km²

Sea surface temperature: largely permanent ice in the long winter as well as short summer seasons

Major currents and gyres: Cape Bathurst polynya

Other oceanographic features: tides less than 2 m; summer sea ice is variable throughout, with a more consistent summer cover in the northern portion than in the south

Physiography: predominantly a system of channels, straits, and fjords surrounding the Arctic islands

Depth: shelf (roughly 0–200 m): 60%; slope (roughly 200–2,500/3,000 m): 40%; abyssal plain (roughly 3,000+ m): 0%. Note that the deepest area, reaching 900 m, is around the Queen Elizabeth Islands.

Major community types and subtypes: estuaries, rocky shores

Productivity: Moderate to low

Species at risk: polar bear, bowhead whale, beluga whale and narwhal

Key habitat: major seabird, waterfowl, and shorebird feeding, staging, and moulting areas

Human activities and impacts: Pesticides used in agriculture in southern and western latitudes are carried by wind to northern latitudes including the Arctic islands; hunting, fishing, adventure tours; oil and gas exploration and recovery; climate change expected to course major ecological impacts.

Air temperatures remain chilly. Even in July, mean daily temperatures average just 10°C. In winter, temperatures average about -30°C, and often much lower.

Biological Setting

Biological productivity in the Central Arctic Archipelago is limited primarily by the availability of sunlight as well as by low year-round cold water temperatures and the very brief summer period. Biological hotspots, consisting of blooms of phytoplankton, occur in spring and summer along the edges of the pack ice and algae grows on the underside of sea ice. These blooms are the basis of some of the Arctic food chains.

During the brief Arctic summer, dozens of species of migrating birds make use of the unpredictable sections of open water that appear in the ecoregion. As the pack ice breaks up, ice edges become vital areas for mammals and seabirds. Taking advantage of the conditions there to feed, stage, and moult are small numbers of tundra swan, loons, geese, ducks, and several species of shorebirds, gulls, jaegers, Arctic tern, alcids, and fulmars.

Polar bears and ringed seals roam throughout the region. Bearded and harp seals are found along the east coast of Ellesmere Island, where open waters promise easy breathing. In winter, the unfrozen North Water Polynya serves as a refuge for marine mammals. In the 19th and early 20th centuries, whalers hunted the bowhead whale almost to extinction. While their numbers have rebounded in western waters, the eastern stock is still severely depleted and the species is considered endangered. Large schools of small Arctic cod exist across the ecoregion, supporting populations of seals, beluga whale and narwhal. The greatest numbers of fish species occur in the west and south, with schools of Arctic cod, Greenland cod, Arctic char, Arctic sculpins, Arctic eelpouts, and snailfish the most common. Other creatures such as anemones, clams, sponges, sea worms, and sea stars are also present.

Human Activities and Impacts

Human presence is sparse in this northern ecoregion, although limited subsistence hunting and fishing, and adventure tours exist. Oil and gas exploration is also found and scientific expeditions have concentrated largely on finding reserves along the edges of the ice pack, but the permanent ice poses formidable challenges to petroleum exploration and drilling.

As with its neighboring ecoregions, this relatively isolated northern system is also affected by substances such as PCBs, DDT and mercury from urban and industrial areas and human activities far to the south. PCBs, for example, are a known contaminant in the breast milk of Inuit women (Miller 2000). Additionally, commercial over-harvesting of mammals and birds has endangered wildlife populations, most notably the bowhead whale (Wiken *et al.* 1996; WHC 2001). Climate change has been altering the ecology of the region and is expected to cause major impacts into the future.

Arctic terns are a ubiquitous species that summers on the coasts of arctic islands. *Photo:* Patricio Robles Gil

The tundra swan is also sometimes called the "whistling swan," not for its bugling call, but for the slow, powerful beating of its wings in flight. *Photo:* Raymond Gehman/National Geographic Image Collection

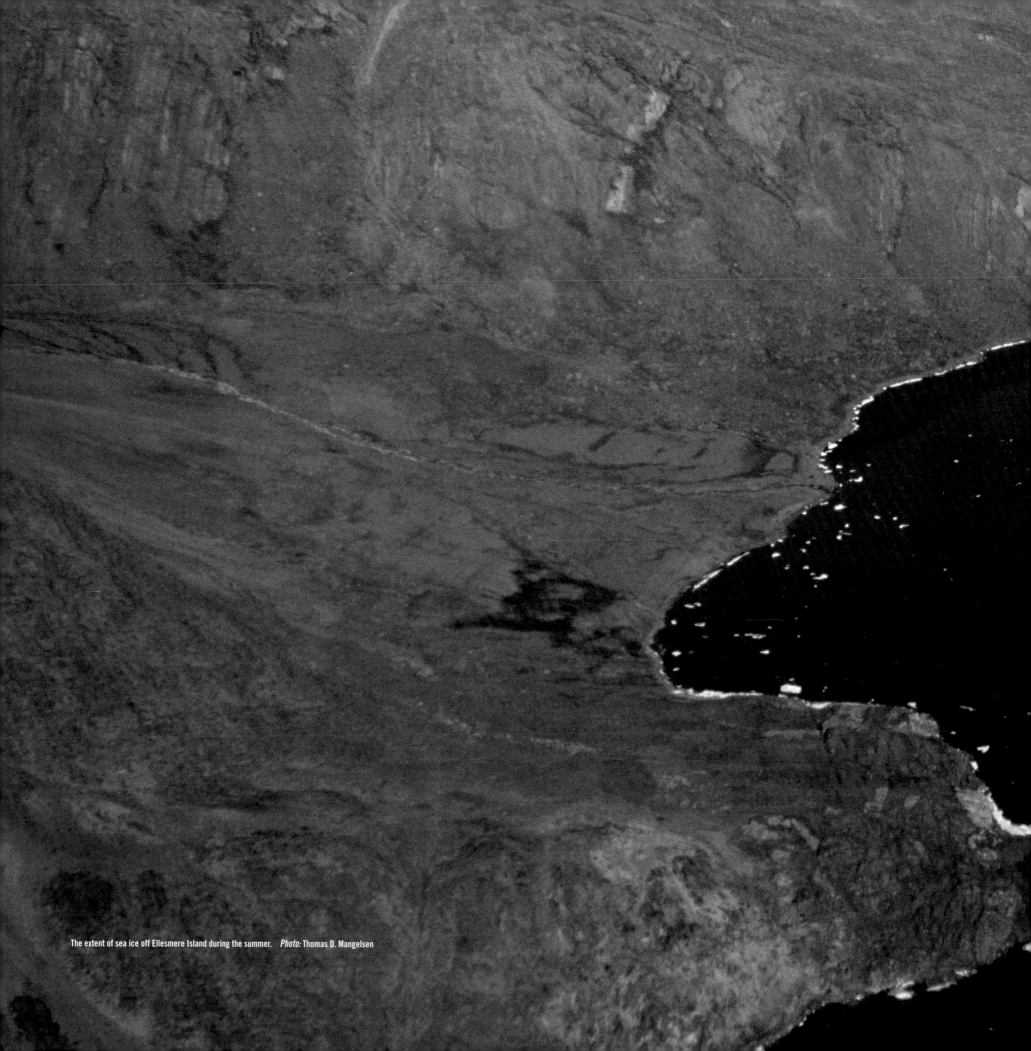

The extent of sea ice off Ellesmere Island during the summer. *Photo:* Thomas D. Mangelsen

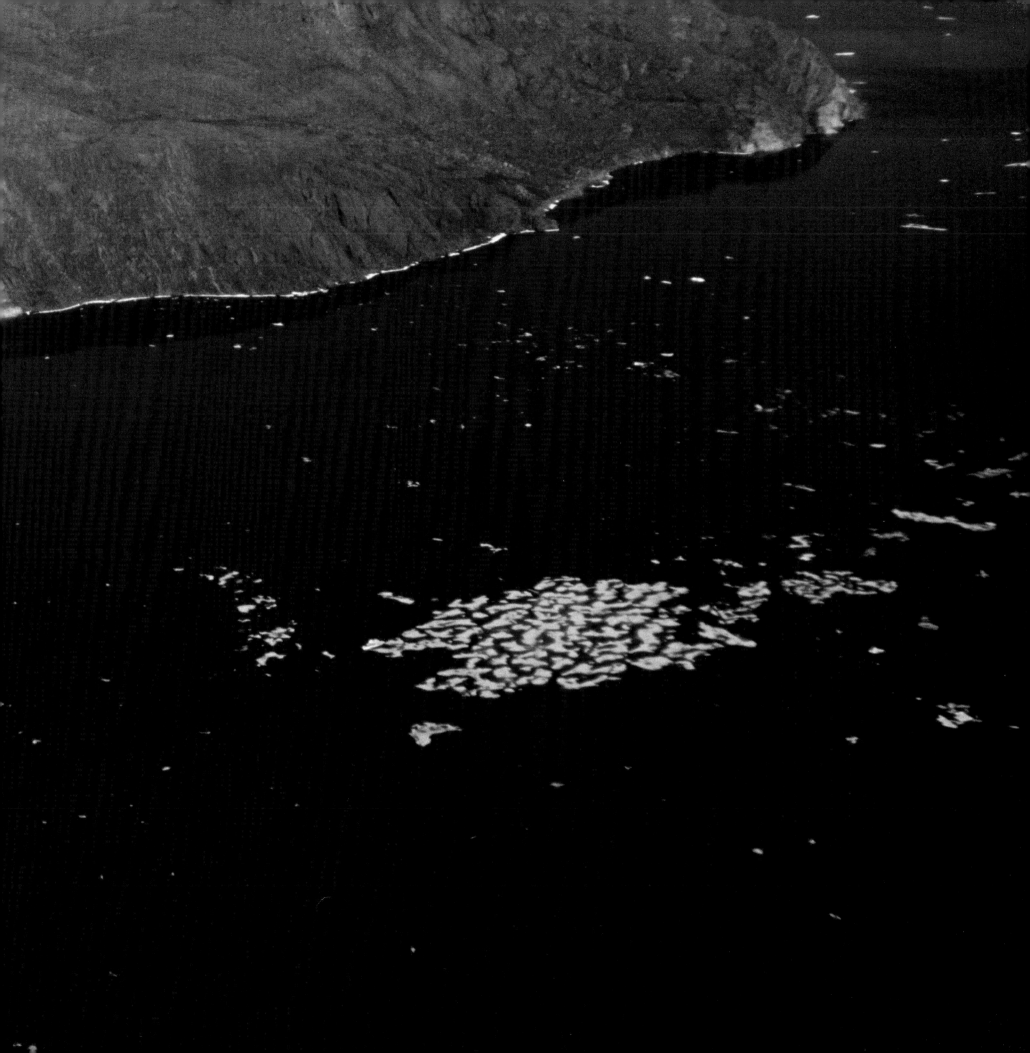

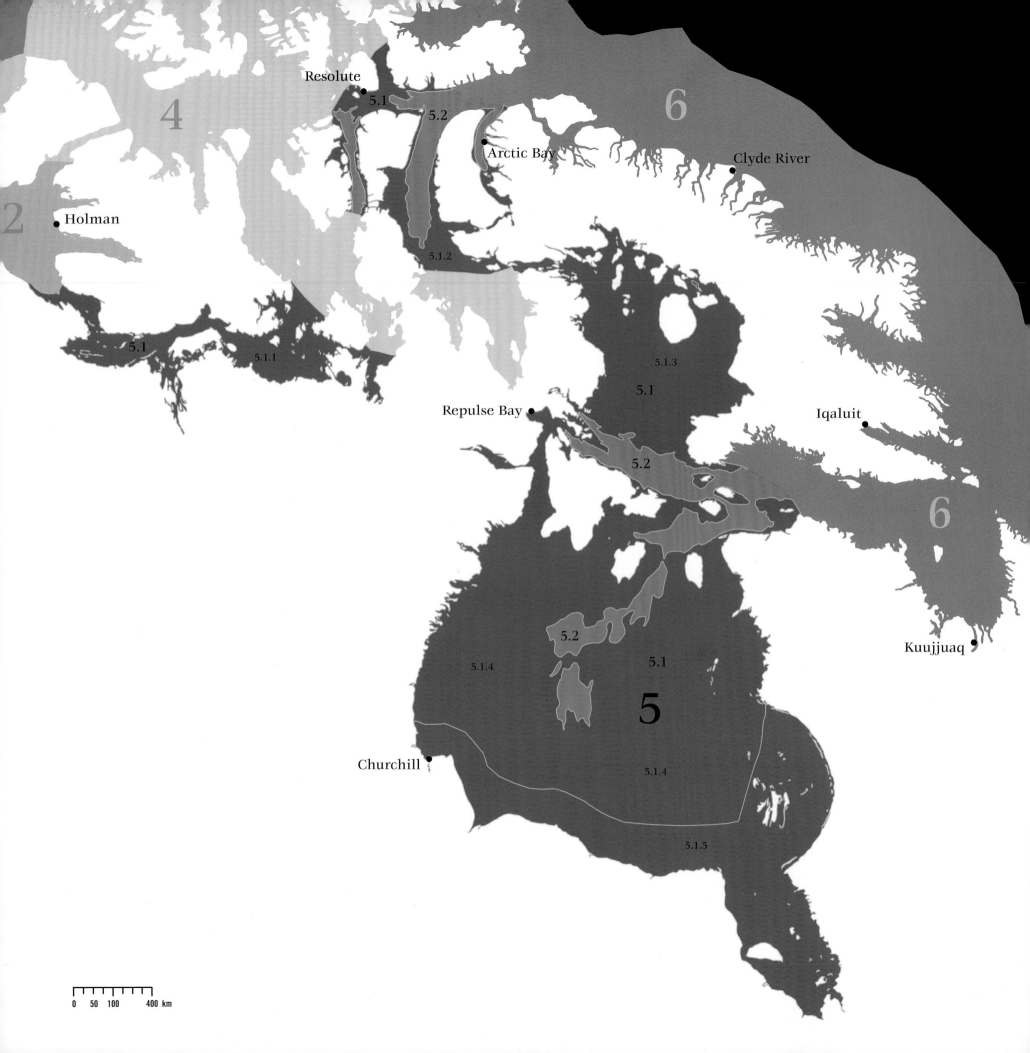

4

6

Resolute
5.1
5.2

Arctic Bay

Clyde River

2

Holman

5.1.2

5.1

5.1.1

5.1.3

5.1

Repulse Bay

Iqaluit

5.2

6

5.2

5.1

Kuujjuaq

5.1.4

5

Churchill

5.1.4

5.1.5

0 50 100   400 km

5. Hudson/Boothian Arctic

Level II seafloor geomorphological regions include:

Level III coastal regions include:

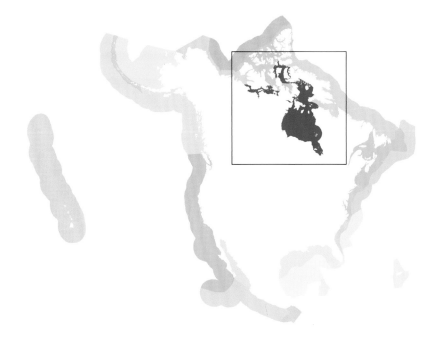

Regional Overview

This is perhaps one of the most unusual marine systems in North America. The primary characteristic of this ecoregion is its Arctic water mass with seasonal ice regimes. Except for the Hudson Bay area, vast and open seascapes are rare in this region. The habitats of this ecoregion are generally composed of a patchwork of interconnecting bays, fjords, channels, straits, sounds, basins, shoals, sills and gulfs. Its northwestern most border includes Prince Regent and Peel Sounds. The ecoregion encompasses part of Lancaster Sound and waters to the west of Baffin Island, including Foxe Basin, most of the Gulf of Boothia (except its southernmost part), Queen Maud and Coronation Gulfs and extends south into Canada's most prominent geographical features, Hudson Bay and James Bay. The ecoregion is one of the richest areas for marine mammals in the world. It is also an important habitat for several species at risk, like the beluga whale and polar bear.

Physical and Oceanographic Setting

The coastline in this ecoregion varies considerably. Precipitous fjords and cliffs are common around the coasts of Baffin Island, Lancaster Sound and Nares Strait, whereas Foxe Basin features flat to rolling coastal plains. Massive glaciers along the backbone of Baffin and Ellesmere Islands often reach from the mountaintops and into the sea. On the eastern side of Ellesmere Island, glaciers extending into the sea calve (break off) huge icebergs into Nares Strait.

During the long winter, sea ice is jammed fast to the land ("landfast ice") and extends over the seas as a solid sheet. Polynyas occur throughout the ecoregion and are critically important to wildlife. The ice cover reaches its maximum thickness in May. In the brief spring and summer periods, the ice breaks up. In the northwestern part of the ecoregion, the sea ice normally shatters into massive sheets that are separated by narrow channels of open water; there, the sea ice persists throughout the summer. In the rest of the ecosystem, the ice is more seasonal. In summer, the massive sheets of ice fracture, drift, and melt away. The process can be very dramatic. Shorelines can be markedly scoured by drifting fragments of ice, or huge ice jams can be driven up on the beaches. As the summer period proceeds, open water can be found further and further north. From year to year, ice conditions are so variable and unpredictable, however, that mariners and navigators of even the most technologically advanced ships can still find themselves blocked. By September, most of the sea ice in the southeasterly parts has either melted or drifted away on southerly currents.

Fact Sheet

Rationale: Arctic water mass and seasonal ice regimes

Surface: 1,294,989 km²

Sea surface temperature: -2–4°C

Major currents and gyres: North Water Polynya in Baffin Bay

Depth: shelf (roughly 0–200 m): 89%; slope (roughly 200–2,500/3,000 m): 11%; abyssal plain (3,000+ m): 0%. Note that water depths of 150–400 m are typical. In Hudson and James Bays the waters are shallow (50–150 m).

Substrate type: rocky to muddy

Major community types and subtypes: estuaries and mud flats

Productivity: Moderately high (150–300 g C/m²/yr), enhanced in coastal waters, near embayments, estuaries, and islands where there is upwelling of nutrient-rich water

Species at risk: beluga whale, polar bear, walrus, narwhal, bowhead whale, peregrine falcon

Key habitat: Polynyas of northern Foxe Basin support high densities of bearded seals and walrus; high-density polar bear denning areas of Southampton Island and Churchill, Manitoba; the region supports most of world's narwhal as well as one-third of North America's beluga—high density along the west coast of Hudson Bay, particularly Nelson River estuary, in summer; whales in high density are found in North Water Polynya in Baffin Bay in winter; adjacent tidal flats and inland marshes are key area for shorebirds and waterfowl

Human activities and impacts: subsistence hunting and fishing, tourism, mining on adjacent lands, hydroelectric development on adjacent rivers, deposition of long-range transported pollutants (e.g., PCBs, DDT, mercury)

Foxe Basin itself is rarely ice-free until late August or early September; open pack ice is common throughout the short summer. Vigorous tidal currents and strong winds keep the pack ice in constant motion and contribute to the numerous polynyas and shore leads which are found throughout the region. This same motion, combined with the high sediment content of the water makes the sea ice of Foxe Basin dark and rough, easily distinguishable from other ice in the Canadian Arctic. In Hudson Bay, the ice cover season is shorter—lasting from October to June. Shore leads along the entire inner edge of the bay are often kept open through the winter by strong prevailing winds, separating the land fast ice along the coast from the pack ice which predominates in most of the bay.

Long periods of daylight in the short summer help stretch the short productive season, but air temperatures remain chilly. Even in July, mean daily air temperatures average just 10°C. In winter, air temperatures average about -30°C, and are often much lower. In the southern range of the ecosystem lie Hudson and James Bays, where the waters are amongst the shallowest (50–150 m) and the climates are the most temperate.

Biological Setting

During the brief Arctic summer, dozens of species of migrating birds make use of the unpredictable sections of open water that appear in the ecosystem. As the pack ice breaks up, ice edges become vital areas for mammals and seabirds. The region provides an example of marine and terrestrial ecosystems that are intricately linked—species like polar bear and Arctic tern, for example, roam between land and sea.

Although relatively little is known about the Hudson Bay/Boothian Ecoregion, recent studies are proving it to be both biologically rich and diverse. The numerous polynyas in the northern Foxe Basin support high densities of bearded seals and the largest walrus herd in Canada (over 6,000 individuals). Ringed seal and polar bear are common, with north Southampton Island and Churchill, Manitoba, being two of the highest-density polar bear denning areas in Canada. This region is also one of the richest in abundance of marine mammals in the world. Most of the world's narwhal population and one-third of North America's belugas, as well as the eastern population of the endangered bowhead whale, spend the summer in these waters; many whales winter in the North Water Polynya in Baffin Bay. Upwards of 20,000 belugas summer along the west coast of Hudson Bay, with the densest concentrations in the larger estuaries, notably that of the Nelson River. Annual autumn concentrations of polar bear along the Cape Churchill coast are also quite exceptional as they gather to await the return of the ice and good feeding. Walrus tend to concentrate around the major polynyas. Some

The Envisat satellite captures an ice-free Foxe Basin, a nearly circular shallow extension north of Hudson Bay, home to the last large landmasses discovered in North America. *Photo:* European Space Agency

20,000 to 50,000 harp seals spend the summer in these waters. This region also has one of the highest population densities of polar bear in the Canadian Arctic.

The Hudson Bay tidal flats and inland marsh areas harbor some of the world's largest concentrations of breeding, moulting and migrating shorebirds and waterfowl. About one-third of eastern Canada's colonial seabirds breed and feed in Lancaster Sound, including more than 700,000 pairs of thick-billed murres, black-legged kittiwakes and northern fulmars. There are also several thousand pairs of black guillemot, Arctic tern, and glaucous, Iceland and ivory gulls; many of these species use the ecoregion to breed. Large colonies of greater snow goose are located in the region as well. Thousands of waterfowl, notably tundra swan, brant, greater white-fronted, Ross's, snow and Canada geese, along with common eider, oldsquaw, and yellow-billed loon breed, moult and stage in the area.

The largest bird concentrations are within the Queen Maud Gulf Migratory Bird Sanctuary, including most of the world's population of Ross's goose. Several shorebird species are also abundant nesters. The region is the main North American stronghold of the Sabine's gull, with some 10,000 pairs nesting here. The Great Plain of the Koukdjuak on Baffin Island is the world's largest goose nesting colony, with upwards of 1.5 million birds, 75 percent of which are lesser snow geese and the remainder Canada geese and brant. Shorebirds and ducks are also abundant, and more than a million cliff-nesting seabirds breed throughout the region, notably thick-billed murres, black-legged kittiwakes and northern fulmars. The region harbors some of the most important North American breeding sites for the Hudsonian godwit and whimbrel, as well as one of the world's largest breeding concentrations of peregrine falcon.

Although some 30 fish species have been reported for the northern portion of the region, one species—Arctic cod—makes up most of the diet of various seabirds and marine mammals. Large schools of this species have been observed in the region. Greenland cod and Arctic char are also common. Some 40 to 50 freshwater, anadromous and Arctic and subarctic marine fish species are found in the waters of the southern portion of the region, with Arctic char, capelin, Arctic cod, Greenland cod, sculpins and blennies being the most abundant.

Human Activities and Impacts

The ecoregion is located mostly in the Nunavut Territory, although it also encompasses parts of the coasts of Quebec,

The little-known, southwestern part of Foxe Basin in the Canadian Arctic supports large concentrations of king eiders, whose males remain in the breeding area for a short time at the start of the season. © Mark Carwardine/OceanwideImages.com

Ontario and Manitoba. Subsistence hunting and fishing for traditional food remain significant human activities in this ecoregion. As mineral prospectors gain more experience in northern latitudes, large mineral deposits may be found, raising questions about mine waste and tailings that could find their way into the sea (MEQAG 1994; Wiken et al.1996). To the south, hydroelectric development will likely continue on the rivers draining into James Bay. Changes in water flow, salt content, and the presence of heavy metals leached from the soil could have unpredictable consequences for the southern portion of the region. The long-range transport of pollutants—such as PCBs, DDT and mercury—from distant places has impacts on the region and adjacent areas as well. Moreover, the possibility that climate change may influence the regimes of ice formation raises concerns about the possibility of commercial shipping routes opening in the future (Wiken et al.1996; WHC 2001), as well as long-term changes to the ecology of the region.

Pod of belugas in the Canadian Arctic. The white adults are easily distinguished from brownish young.   *Photo:* Art Wolfe

6. Baffin/Labradoran Arctic

Level II seafloor geomorphological regions include:

6.1	Baffinian Shelf
6.2	Ungava/Labradoran Shelf
6.3	Grand Banks
6.4	Hudson Trough
6.5	Baffin/Labradoran Slope
6.6	Labrador Plain

Level III coastal regions include:

6.1.1	Baffin Estuarine Areas
6.1.2	Baffinian Neritic
6.2.1	Labrador Estuarine Areas
6.2.2	Ungava/Outer Banks/Labradoran Neritic
6.3.1	Grand Banks Neritic

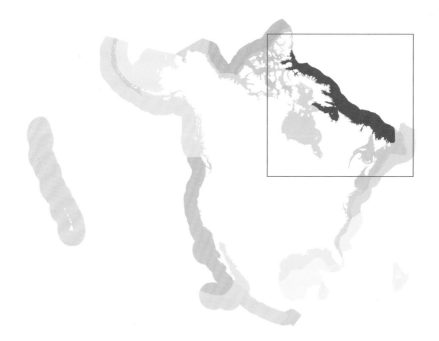

Regional Overview

Significant proportions of the North American and world populations of seabirds live in the Baffin/Labradoran Arctic Ecoregion. Steep, rocky cliffs and thousands of islands provide ideal habitat for some of the largest seabird colonies in eastern North America. The region extends from roughly the middle of Ellesmere Island, along the eastern edge of Baffin Island, and onto the coast of Labrador and northern Newfoundland. It forms a transition between the cold northern waters and the more temperate southern waters. Sea ice is common throughout much of the region, depending on the season and latitude.

Physical and Oceanographic Setting

The ecoregion is in fact subarctic in nature as a result of the influx of the warmer waters from the West Greenland Current. The Labrador Current—the main current in the region—flows south to meet the Gulf Stream in the more southern Canadian Atlantic. The region is characterized by seasonal ice that begins to form off the coasts of Ellesmere and Baffin Islands as well as Labrador in November or December; by February or March, it usually reaches the northeast coast of Newfoundland, accompanied by thousands of icebergs.

Fjords, cliffs, and rocky areas characterize much of the immense coastline contained in this region. Its continental shelf extends 50–150 km from shore, at a depth of 200–300 m.

Biological Setting

The region is home to a great number of marine mammals, fish and birds. Humpback, sei, fin, bowhead, beluga and minke whales are most likely to be spotted, along with the occasional blue whale. Millions of harp seal and hooded seal breed and migrate along the coasts and others, such as the bearded, ringed and harbor seal, also occur.

Significant proportions of the North American and world populations of seabirds live in the region. Steep, rocky cliffs and thousands of islands provide ideal habitat for some of the largest seabird colonies in eastern North America, notably razorbills, common murres, and northern fulmars, thick-billed murres and northern gannets. Millions of gulls, murres, dovekies, fulmars, sea ducks and shearwaters winter offshore. There are over 400 seabird colonies along the coast of Labrador, of which seven are considered "major" (i.e., more than 500 breeding pairs).

Fact Sheet

Rationale: transitional region between northern cold waters and more temperate southern waters; seasonal ice

Surface: 1,449,632 km²

Sea surface temperature: August surface temperatures vary between 3–19°C

Major currents and gyres: Labrador Current and West Greenland Current

Depth: shelf (roughly 0–200 m): 53%; slope (roughly 200–2,500/3,000 m): 35%; abyssal plain (roughly 3,000+m): 11%

Substrate type: rather rocky and barren with talus that skirts coastal cliffs that rise steeply from the sea

Major community types and subtypes: Although the intertidal zone is quite barren due to the scouring action of sea ice, the subtidal benthic community is rich. Fish diversity is low, and Arctic cod is dominant. Other species that are found in the region include bowhead, northern bottlenose, sperm, blue, fin, sei, minke, humpback, pilot, killer and beluga whales; narwhal; harp, hooded and ringed seals; walrus; polar bears; thick-billed and common murres; razorbills; King eiders; Atlantic puffins

Productivity: moderately high (150–300 g C/m²/yr)

Species at risk: Atlantic cod; blue, beluga, fin, right and humpback whales; leatherback sea turtle

Key habitat: Steep, rocky cliffs and thousands of islands provide ideal habitat for some of the largest seabird colonies in eastern North America

Human activities and impacts: fishing, tourism, mineral mining, oil and gas exploration, shipping.

Abundant fish species in this ecosystem include: Atlantic cod, Arctic cod, Arctic char, Arctic sculpins, Greenland halibut, and Greenland cod, Atlantic herring, Atlantic halibut, redfish, American plaice, haddock, silver hake, pollock, northern flounders, and Atlantic mackerel, Atlantic salmon and capelin.

Human Activities and Impacts

Fishing has been one of the main human influences in this region, where Canada's cod stocks have been nearly depleted. In 1992, the government of Canada imposed a moratorium on cod fishing in most of Canada's Atlantic waters in hopes the stocks would rebuild. More than 10 years later, stocks are not recovering, at least not at the rate expected (COSEWIC 2003).

Concern has been voiced about the exploitation of giant mineral deposits such as those at Voisey's Bay on the northern Labrador coast, increased activity related to oil and gas exploration, and increased levels of shipping to supply communities. Cruise ships also pose a concern for the region.

Some of North America's largest northern gannet colonies are found at Cape St. Mary's in Newfoundland and Labrador and on Quebec's Gaspé Penninsula.   *Photo:* Alan D. Wilson/www.naturespicsonline.comg

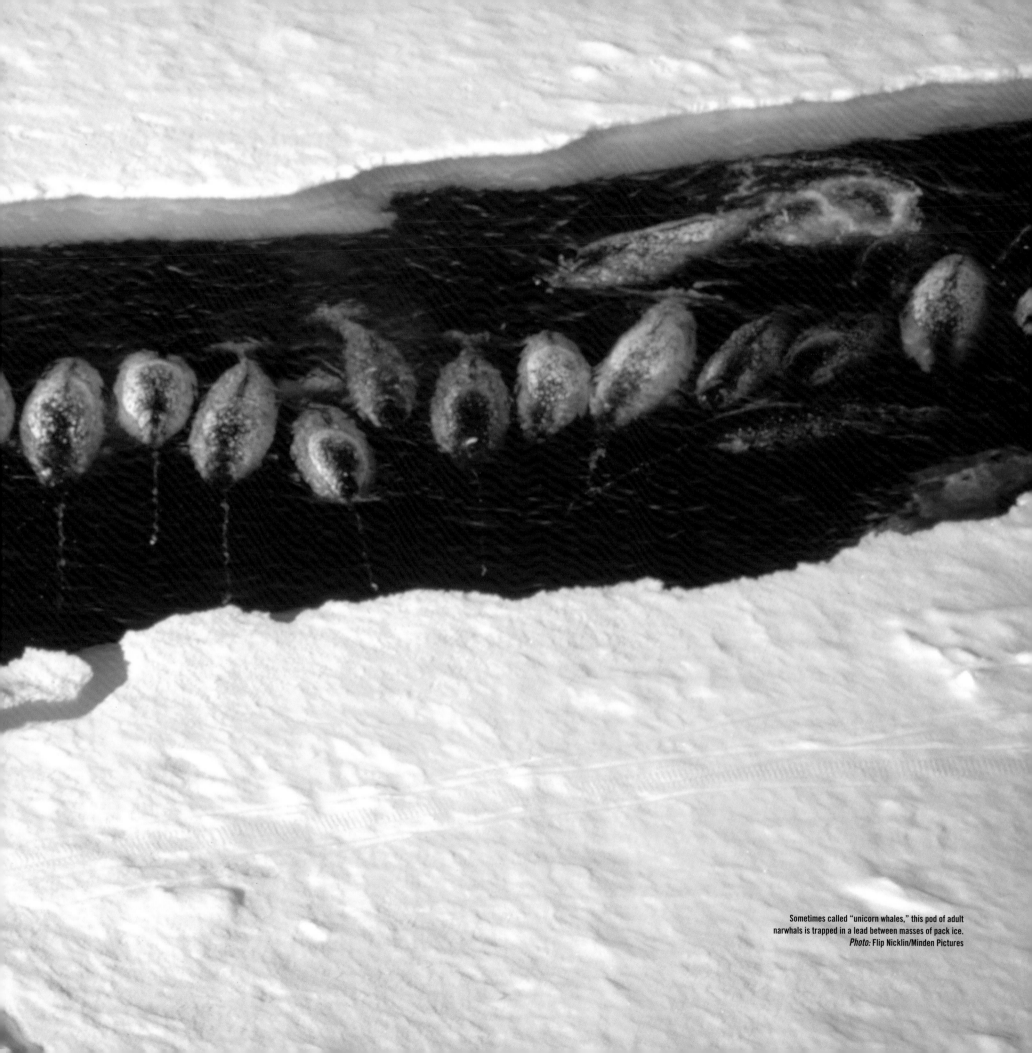

Sometimes called "unicorn whales," this pod of adult narwhals is trapped in a lead between masses of pack ice.
Photo: Flip Nicklin/Minden Pictures

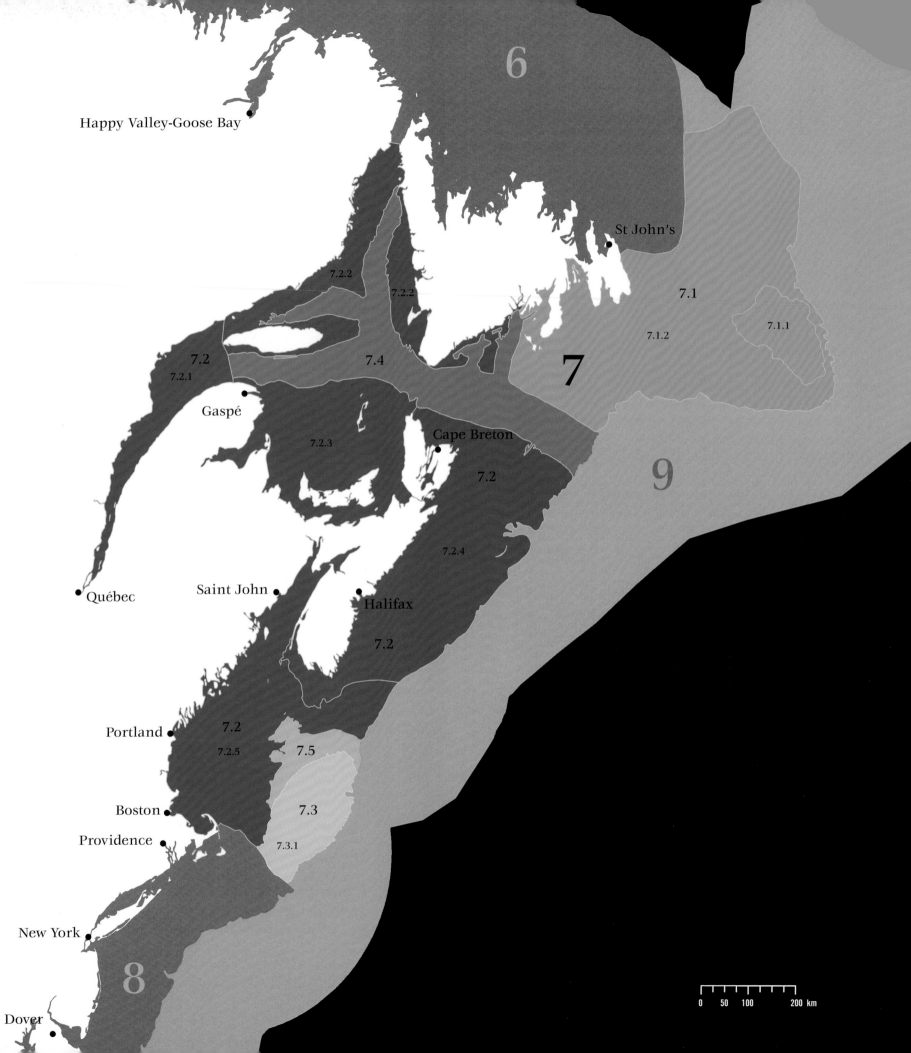

7. Acadian Atlantic

Level II seafloor geomorphological regions include:

7.1	Grand Banks
7.2	Acadian Shelf
7.3	Georges Bank
7.4	Laurentian/Esquiman Channel
7.5	Northeast Channel/Georges Basin

Level III coastal regions include:

7.1.1	Southeast Shoal
7.1.2	Grand Banks Neritic
7.2.1	St. Lawrence Estuarine Area
7.2.2	North Gulf Neritic
7.2.3	Magdalen Shallows
7.2.4	Scotian Neritic
7.2.5	Gulf of Maine/Bay of Fundy
7.3.1	Georges Bank Neritic

Regional Overview

The Acadian Atlantic Region—one of the most productive marine areas in the world—is also known for its valuable and highly depleted fish communities, extreme tides, and rich mineral deposits (oil and gas). It is also downstream and adjacent to some of the largest centers of urban and industrial development in the northeast of North America. The Acadian Atlantic Region extends from Newfoundland and the Grand Banks, south to the Scotian Shelf and on to Cape Cod. Cape Cod forms a partial barrier to oceanic currents and many species; the southern boundary of this region extends from Cape Cod, along the Great South Channel south of Georges Bank to the shelf edge. On its seaward boundary, the Acadian Atlantic Region borders the offshore zone that is influenced by the Gulf Stream (Northern Gulf Stream Transition Ecoregion). The area encompasses a coastline formation heavily influenced by glacial processes, resulting in complex geomorphology, cliffs, rocky coastal zones and exposed resistant bedrock. Numerous relatively small coastal watersheds deliver fresh water to important estuaries, pocket fresh and salt marshes that support many coastal fisheries and life cycles of other species.

Physical and Oceanographic Setting

The Acadian Atlantic Region is characterized by a broad continental shelf extending up to 200 km from the coastline in its northern extent, narrowing to less than 100 km at Cape Cod. The shelf is also relatively shallow in this region, often occurring at 100–150 m depth in some places. The continental slope is marked by numerous steep canyons cutting from the shelf edge to a plateau region at 400–1,000 m, from where it slopes to an abyssal plane more than 4,000 m deep. The coastline of the region is characterized by terminal glacial moraines, such as at Cape Cod, to rocky shoreline, such as coastal Maine and Nova Scotia, as well as extensive cliffs along Newfoundland. The bathymetry around the Gulf of Maine and Bay of Fundy is extremely complex, marked by an elaborate system of banks and canyons.

Due to the numerous large coastal watersheds and estuaries in this region, coastal waters of the area are affected by spring flood and the summer wet season. In the Gulf of St. Lawrence, fresh water flowing from the St. Lawrence River keeps the salt levels low through most of the year. Tidal ranges within the region are normally 1 to 2 m, though in parts of the Gulf of St. Lawrence, tidal action is small, with average fluctuations of less than a meter.

Fact Sheet

Rationale: The region is defined by current regime, physiography and cold sea surface temperature. Cold, low salinity water is transported in the Labrador Current from the Arctic Ocean southward into the Gulf of St. Lawrence and the Gulf of Maine. Distribution of many species breaks at southern boundary at Cape Cod. Its eastern boundary is at the shelf break, and its northern boundary at the permanent ice line north of Labrador.[9]

Surface: 823,991 km²

Sea surface temperature: -1–17.5°C (winter), 10–23°C (summer)

Major currents and gyres: Labrador Current, West Greenland Current; important pathway in the Gulf of Maine transports the western part of Labrador Current north to south with exits through the Great South Channel; upwellings occur around Georges Bank and the Flemish Cap

Other oceanographic features: high diurnal tide range; strong frontal passages in winter; partially ice-covered in winter

Physiography: extremely complex, crenellated coastline; broad shelf marked by shoals, steep channels, deep trenches and numerous banks (Grand Banks, Flemish Cap, Georges Bank, Brown's Bank)

Depth: shelf (roughly 0–200 m): 86%; slope (roughly 200–2,500/3,000 m): 14%; abyssal plain (roughly 3,000+ m): 0%. Note that shelf was calculated from shore to shelf break at 500 m in northern region around Newfoundland and to 200 m shelf break in southern extent.

Substrate type: silts, cobble, gravel, and resistant rock

Major community types and subtypes: Characteristic biological communities of rocky coastal zones, estuaries, salt marshes, tidal flats, sandy beaches, shoal, deep-sea, slope communities

Productivity: moderately high productivity ecosystem (150–300 g C/m²/yr)

Species at risk: North Atlantic right, blue, fin, sei and humpback whales; beluga; leatherback sea turtle; Atlantic and shortnose sturgeons; barndoor skate; and sand tiger shark

Important introduced and invasive species: compound sea squirt. In the wetland environments, purple loosestrife is an invasive species

Key habitat: Pocket salt marshes along New England coast; important region for estuarine dependent species in Cape Cod, Boothbay Harbor; whale feeding habitat along shelf break; Atlantic cod and Georges Bank yellowtail flounder habitat in shallows around upwellings at Georges Bank

Human activities and impacts: coastal development, especially around urban areas, fishing, ocean aquaculture, tourism, commercial shipping and navigation

---

9 On the basis of bathymetry, hydrography and productivity, this region has been partitioned into three separate Large Marine Ecosystems (LMEs): the southern end of the Newfoundland-Labrador Shelf LME, the Scotian Shelf LME, and the northern half of the Northeast US Continental Shelf LME (Zwanenburg *et al.* 2002). The Acadian Region may therefore represent a heterogeneous region and for certain purposes, the LME delineation may be more useful.

The Bay of Fundy is a significant exception, with a tidal range of over 16 m—one of the largest in the world.

Ocean currents generally flow from north to south, parallel to the coast, with an onshore component in the New England Bight during the winter months. During summer, the same north-south flow predominates in the southern range of the region, while from New England up through Canadian waters the flow reverses to a northward direction. Off Newfoundland, mixing of the warm currents from the south and the Labrador Current from the north produces some of the most famously dense fog banks on the planet. As the Labrador Current reaches the northern extreme of Georges Bank, it splits into two, with a major portion flowing east around the bank, and another part flowing west of the bank into the Gulf of Maine and out the Great South Channel. The outer portion of the region is temperate due to the moderating influence of the massive Gulf Stream nearby, while the Gulf of St. Lawrence is generally more subarctic, under the influence of the Labrador Current. Prevailing winds blowing off the land from the west and southwest also moderate the ocean climate. In August, surface water temperatures can vary between 10 and 23°C.

Around April, ice covered areas begin to clear and by July most coasts are free of ice. In winter and early spring, ice can also be plentiful along the east coast of the Avalon Peninsula (in Newfoundland) and in the Cabot Strait between Newfoundland and Nova Scotia. The Gulf of St. Lawrence is predominantly ice-covered for several months in winter. Icebergs are a common sight in late winter, spring and even early summer off the Newfoundland coast and on the Grand Banks. From the outer coast of Nova Scotia and the Bay of Fundy down to Cape Cod, the region is essentially free of ice.

Biological Setting

The waters over the continental shelf in this region are famous for their prolific communities of marine mammals, birds and fish. Benthic communities in the region are diverse, supporting a variety of marine plants, such as seaweeds, kelps, brown, red, and green algae, and are rich with invertebrates such as barnacles, sea stars, crabs, American lobster, sponges, scallops, clams and jellyfish.

Twenty-two species of cetaceans and six species of seals occur in the ecosystem. Humpback, fin and minke whales are likely to be spotted in nearshore waters; in addition, sei, bowhead, and beluga whales can also be found. The region's coastal waters are home to the highly endangered North Atlantic right whale as well. Sperm and blue whales can be found further offshore, although blue whales are also found in the Gulf of St. Lawrence and on the shelf. Porpoises and dolphins also commonly occur. Tens of thousands of harp seals as well as thousands of hooded seals breed on the ice in the Gulf of St. Lawrence, while gray and harbor seals are resident throughout much of the region. Key habitat for many cetaceans is also found in the region: a relatively small area in the Great South Channel is important habitat for the fin whale; major feeding grounds for the humpback whale occur throughout the greater Gulf of Maine (particularly the Great South Channel, Stellwagen Bank and Jeffreys Ledge), east and south coasts of Newfoundland, waters off southeastern

Labrador, on the edges of the Grand Banks, and in the Gulf of St. Lawrence; feeding and nursery areas for the North Atlantic right whale in the region include Cape Cod Bay/Massachusetts Bay, Great South Channel, Stellwagen Bank, Jeffreys Ledge, the Bay of Fundy, Brown's Bank and Roseway Basin; and the Gulf of St. Lawrence, the Scotian Shelf, and waters off southern Newfoundland are important to the blue whale.

During the summer, the steep, rocky cliffs and thousands of islands provide ideal habitat for some of the largest seabird colonies in the world. Around Newfoundland, in particular, there are over 10,000,000 nesting seabirds in the summer. Witless Bay supports the largest breeding population of Atlantic puffins in North America. With 3.3 million pairs, Baccalieu Island contains the world's largest nesting population of Leach's storm-petrels. Historic Funk Island hosts the single-largest breeding aggregation of common murres in the world. Millions of seabirds, mainly shearwaters and alcids, winter in the offshore waters of Newfoundland. Seal Island in the Bay of Fundy supports significant colonies of reintroduced puffins as well as Arctic tern. Colonies of northern gannet, Atlantic puffin, Leach's storm-petrel, double-crested cormorant, and several varieties of gull, including the great black-backed gull, also nest in the Gulf of St. Lawrence on Anticosti Island, Bonaventure Island, and the Magdalen Islands. The region also hosts northern fulmars, thick-billed murres, kittiwakes, eiders and cormorants.

The region is a particularly valuable area for fish. Commercially important fish species include Atlantic cod, redfish, Atlantic herring, silver hake, and Greenland halibut (or turbot). The Atlantic cod and haddock fisheries, concentrated on Georges Bank and Grand Banks as well as nearshore New England, depend on spawning areas in the New York Bight and on Georges Bank itself. These fisheries, at risk due to commercial fishing pressure, are tightly regulated. The Atlantic herring has major summering grounds in the New England Bight and off Nova Scotia, migrating to important spawning areas on and near Georges Banks, then south to wintering grounds on the continental shelf offshore from Chesapeake Bay. Atlantic mackerel migrate to and from the shelf edge and slope to coastal waters. Bluefin tuna are harvested in large concentrations of fish larger than 150 kilograms offshore of New York, New England and the Atlantic provinces. Their spring migration brings this species from southern Florida along the margin of the continental slope.

In the northern part of the region, one can find one of the most biologically productive marine areas in the world—the Grand

Colorful puffins dwell on rugged sea cliffs and feed in cold oceanic waters. *Photo:* Patricio Robles Gil

Picturesque rocky coastline and tidal ponds near Thunder Hole, Acadia National Park, Maine.   *Photo:* Tim Fitzharris/Minden Pictures

Banks. The confluence of the Labrador Current and the Gulf Stream and the tidal mixing of the water column on the shallows of the continental shelf provide ideal feeding and spawning conditions for thousands of species.

The Magdalen Shallows are an area of high productivity for fish, and important populations of northern shrimp and snow crab are also found here. American lobster and scallops are found in the nearshore area, while shrimp occur in deeper waters. The Magdalen Shallows are part of an area where the composition of the marine fauna is a balance of northern species (with cold water subarctic affinities) and southern species.

The low-lying beaches, salt marshes and tidal flats of the coastal marshes found in Narragansett Bay, Buzzards Bay, Waquoit Bay, Cape Cod Bay, the Upper Bay of Fundy and the southern Gulf of St. Lawrence are dominated by burrowing organisms such as corophium and annelid worms. These are extremely abundant at or just below the surface of the mud flats and are fed upon by millions of shorebirds.

An extremely complex bathymetry, composed of sills, rises, ridges and basins, characterizes the Gulf of Maine and Bay of Fundy in the area relatively close to shore. This habitat is formed in part by the huge tidal fluctuations in the Upper Bay of Fundy. These areas are extremely important fish, shellfish, and mammal habitats. The rich, deep water forces up and over the shallow bank system, causing an upwelling of nutrient-rich bottom waters. The result is a rich, multi-layered and productive food web supporting one of the most important whale feeding grounds in the world.

The estuaries of the Gulf of Maine are thought to be vital to nearly three-quarters of the commercially significant fish species in the area. They serve as nursery areas for juvenile fish and the planktonic larvae of mollusks, crustaceans, and other invertebrates. Phytoplankton blooms, which can turn the water green with life every spring, are the first link in the food chain of the Acadian Atlantic Region. Marine plants such as seaweeds and kelp are prolific, especially in intertidal zones. Extensive salt marshes occur throughout these estuaries, particularly in New Brunswick, Nova Scotia, and Prince Edward Island, but less frequently in Newfoundland, Labrador and New England. These tidal wetlands are home to the highly salt-resistant marsh grass, such as saltmeadow cordgrass, as well as a variety of other plants, including marsh spikegrass, wild barley, sea lavenders and sea plantains.

Human Activities and Impacts

Fishing has provided the way of life for Atlantic coastal communities for hundreds of years. Technological advances, increased demand and over-capitalization have contributed to the collapse of stocks of species such as cod. Fishing for species of lobster, shrimp, and crab still provides a livelihood for some families. Aquaculture is increasing, including salmon, scallop, and cod farming under way in New Brunswick, Nova Scotia, Newfoundland and New England. Commercial harvests of many of these species are no longer sustainable, and 10 out of 29 major stocks in US waters of the region are overfished (NMFS 2004). Habitat damage from fishing gear and loss of nursery habitat have also affected the sustainability of these stocks. There are fears the once-rich Atlantic cod stocks may never again support a commercial fishery (Oceans Ltd. 1994), however, certain other species (e.g., Georges Bank yellowtail, haddock and scallops) appear to be recovering.

Shipping represents one of the primary commercial activities throughout the region, providing livelihoods to millions. Economic development in Newfoundland is strongly linked to offshore oil and gas exploration and production in the rich Hibernia, Terra Nova, White Rose and Hebron oil fields. Potential exploitation of these, as well as other important reserves off Nova Scotia along the Scotian Shelf, along with expanded exploration and shipping, further increase environmental risks in the region.

The dramatically increasing population density in adjacent coastal areas and related coastal development, municipal sewage, agriculture, and industrial development also threaten the region's ecosystems. Urban sprawl and industrial development around coastal cities and along the banks of the St. Lawrence River (and even as far as the Great Lakes), Cape Cod, and the New England coastal bays such as Buzzards and Waquoit Bays has destroyed much of the wildlife habitat. Toxic chemicals from these areas are having a substantial effect on the region's marine life, particularly on species higher up the food chain, such as the endangered beluga whale of the Gulf of St. Lawrence. Climate change and variability further threaten the ecology of the region, potentially compounding other threats for some species.

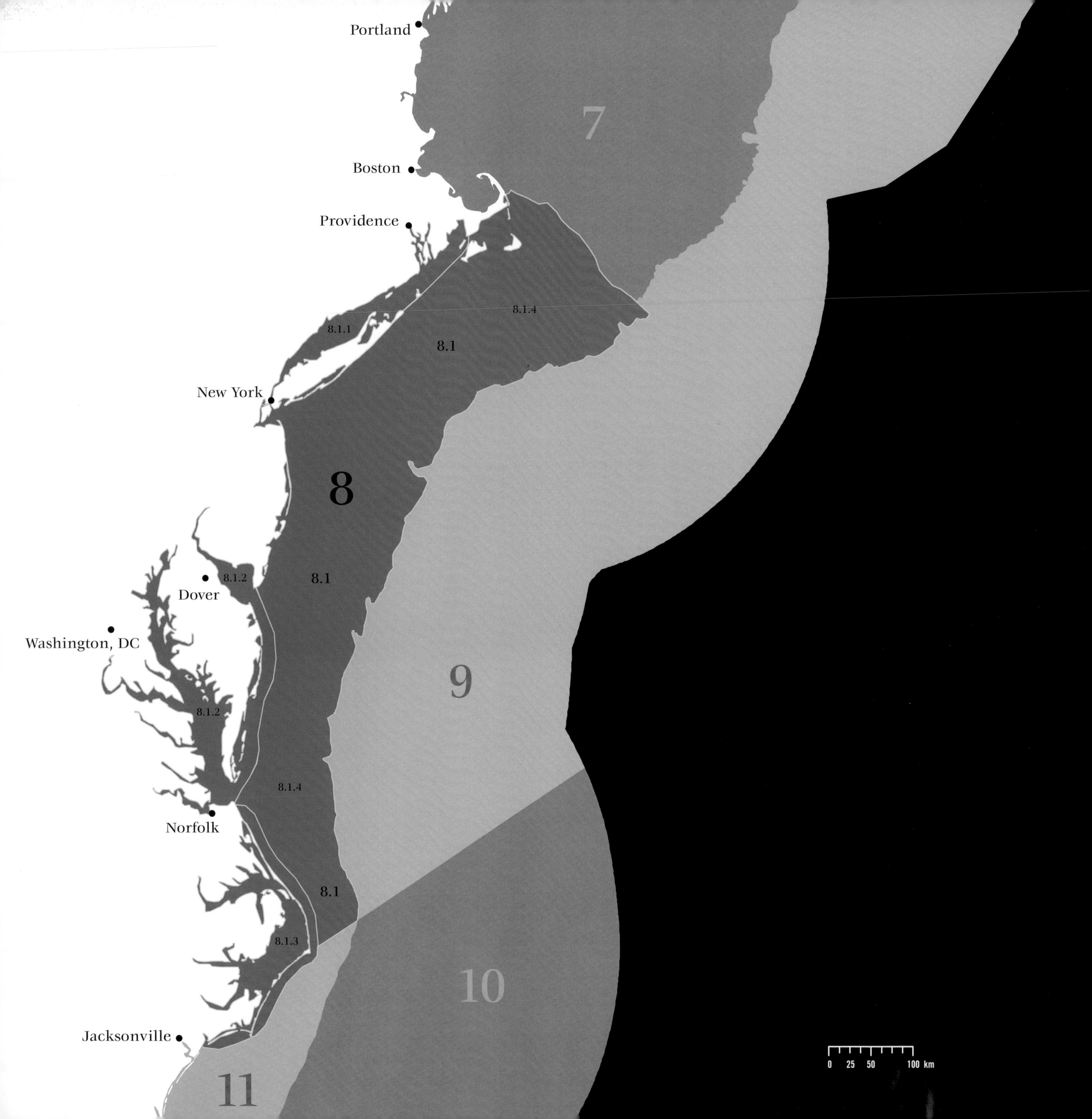

Portland

Boston

Providence

8.1.4

8.1.1

8.1

New York

8

7

9

8.1

8.1.2

Dover

Washington, DC

8.1.2

8.1.4

Norfolk

8.1

8.1.3

10

Jacksonville

11

0   25   50   100 km

8. Virginian Atlantic

Level II seafloor geomorphological regions include:

   8.1   Virginian Shelf

Level III coastal regions include:

   8.1.1   Long Island Sound/Buzzards Bay
   8.1.2   Delaware/Chesapeake Bay
   8.1.3   Pamlico Sound/Outer Banks
   8.1.4   New York Bight

Regional Overview

The Virginian Atlantic Region extends along the eastern North American continent from Cape Hatteras northward to Cape Cod, lying within the temperate climatic zone, and interposed between the east coast and the Northern Gulf Stream Transition Region offshore. It supports key ecological assemblages and commercially important fisheries with binational (Canada-US) ranges. Chesapeake Bay, the largest estuary in the United States and one of the largest in the world, also lies within this region. The region is home to a historically enormous oyster fishery that has dwindled in recent years due to pollution, overfishing and disease. The region's coastlines were formed by glacial processes and by river sedimentation, resulting in complex and variable geomorphology. To the north, the coastlines are typically rocky, resistant bedrock formations and small marsh areas. To the south, bar-built estuaries, barrier islands and drowned river valleys predominate, although rocky coastal zones are found as well. Numerous coastal watersheds deliver fresh water to important estuaries and fairly extensive fresh brackish and salt marsh systems.

Physical and Oceanographic Setting

The Virginian Atlantic Region has a broad continental shelf extending as much as 150 km from the coastline north of Long Island, narrowing to about 40 km at Cape Hatteras. As in the Acadian Region to the north, deep canyons cut through the continental slope, particularly off the coasts of New York and New Jersey. The coastline of the Virginian Atlantic is characterized by barrier island systems, such as Pamlico Sound, drowned river valleys, such as Chesapeake Bay and Delaware Bay, terminal glacial moraines such as Long Island and southern Cape Cod, and is interspersed by high energy rocky shoreline and small to extensive coastal marshes. Due to significant, large coastal watersheds and estuaries in this region (Hudson River, Delaware Bay, Chesapeake Bay, Pamlico Sound), several areas experience freshening of the nearshore water mass during spring flood and the summer wet season. The Chesapeake Bay includes eleven major tributaries delivering water into the bay and coastal zone on a seasonal basis. The Chowan, Roanoke, Pamlico and Neuse Rivers supply up to 424.5 cubic meters per second that empty into the Pamlico-Albemarle Sound, North Carolina. Ocean currents generally flow from north to south parallel to the coast year-round. The eastward-turning Gulf Stream, with its moderating influence on the climate

Fact Sheet

Rationale: defined by sea surface temperature and currents; extends from Cape Hatteras in the south to Cape Cod in the north, where coastal physiography and Georges Bank somewhat restrict the exchange with the Gulf of Maine. Shelf waters are insulated from the Gulf Stream by the adjacent deep waters of the Gulf Stream transition region beyond the shelf break.

Surface: 150,027 km²

Sea surface temperature: 2–20°C (winter), 15–27°C (summer).

Major currents and gyres: western edge of adjacent Gulf Stream creates upwellings along shelf break in this region

Other oceanographic features: strong stratification along coast where estuaries of fresh water occurs; export from Delaware, Chesapeake Bays

Depth: shelf (roughly 0–200 m): 100%; slope (roughly 200–2,500/3,000 m): 0%; abyssal plain (roughly 3,000+ m): 0%

Major community types and subtypes: rocky coastal zones, rocky headlands, estuaries, salt marshes, tidal flats, sandy beaches, barrier island systems

Productivity: High (>300 g C/m²/yr), one of the most productive in the world

Species at risk: North Atlantic right whale, fin whale, leatherback sea turtle, diamondback terrapin, shortnose sturgeon, Atlantic sturgeon, barndoor skate, night shark, dusky shark, and sand tiger shark

Important introduced and invasive species: compound sea squirt, veined rapa whelk

Human activities and impacts: high turbidity and nutrient input from urbanization and agro-development in watersheds has impacted the ecology of many nearshore areas, including most of the major estuaries in the region. Centuries of human development destroying spawning habitat has been a problem for the anadromous sturgeons and other species. The Atlantic sturgeon population has declined due to commercial fishing, and the shortnose sturgeon has declined largely as a result of bycatch. Dam construction has eliminated habitat for most anadromous species. Commercial shipping and navigation are prevalent in the region.

of the Virginian Atlantic Region, lies just offshore, creating a zone of transition to the east where complex current structures lead to upwellings.

Biological Setting

Coastal communities of the region include numerous relatively small but ecologically important tidal salt marshes and several large and significant estuaries. Both wetlands and estuaries are important in supporting coastal fisheries as spawning and nursery grounds, particularly in Chesapeake Bay, Delaware Bay and Narragansett Bay. The region is highly productive. Important fisheries include the Atlantic mackerel, which migrate between the shelf edge and slope to coastal waters of the New York Bight. Bluefin tuna larger than 150 kilograms are harvested in large quantity offshore of New York. Spring migrations of tuna bring this species and bigeye tuna from southern Florida to the Virginian Atlantic along the margin of the continental slope. The important Atlantic cod and haddock fisheries depend on the region's spawning areas found in the New York Bight. These fisheries are endangered due to commercial fishing pressure and are highly regulated.

The Atlantic herring has major wintering grounds on the continental shelf offshore of Chesapeake Bay, migrating northward during summer. American lobster is found in major commercial concentrations along the coast from north of Delaware Bay and in the New York Bight and Long Island Sound and in a similar distribution along the continental slope. The American shad has a complex life history and migrates from throughout the offshore Atlantic to estuarine spawning grounds in Pamlico Sound, Chesapeake Bay, Delaware Bay, Long Island Sound. The eastern oyster is found in large concentrations in Chesapeake Bay, Delaware Bay, the north shore of Long Island Sound, and Peconic Bay.

The estuaries of the region are crucial for other species as well. The striped bass, with a range extending from the Gulf of Mexico and Florida to the Maritimes of Canada, is one of the premier

Coastal zones like Chesapeake Bay are important breeding areas and habitats for many marine species, but they often suffer from the impacts of human activities. *Photo:* Jim Wark

recreational and commercial species along the east coast. Atlantic coastal fisheries for striped bass rely primarily on populations that spawn in the Hudson and Delaware estuaries and in tributaries of Chesapeake Bay. Chesapeake Bay has historically produced most of the striped bass found along the coast. Likewise, the nursery area offered by southern Chesapeake Bay, Pamlico Sound and Delaware Bay for the (Atlantic) black sea bass is critical to a fishery that extends from southern Florida to Cape Cod. Though the blue crab fishery continues, it is endangered, tightly regulated, and diminished relative to historical harvest yields. Delaware Bay is an important area of annual migration for the horseshoe crab. Wetlands in coastal areas and estuaries continue to be lost to development, and critical areas within estuaries such as shell reefs are often lost to dredging.

Human Activities and Impacts

The overall condition of estuaries in the Virginian Atlantic Region is poor (EPA 2005). Problems associated with excess nutrients from human activities and low oxygen levels in bottom waters are generally more severe than in the Acadian Atlantic Region to the north (EPA 2005). The region includes the largest population centers on the East Coast, with extensive urban and industrial development. Historically enormous oyster fisheries have dwindled in recent years due to diseases such as viruses Dermo and MSX, overfishing, and pollution. Bioaccumulation of heavy metals, pesticides and toxic shellfish bacteria have depleted the oyster fishery in some areas. For example, a previously thriving economy based on the Chesapeake Bay oyster fishery has dwindled since the early 1970s, severely impacting some communities in coastal Maryland and Virginia. Many oyster areas have been closed to harvest as rejuvenation of the fishery is attempted. The Virginia blue crab fishery in southern Chesapeake Bay has also declined in recent years, more due to overfishing and harvesting of sub-spawning age individuals than to pollution. In federal waters, two out of 11 major fish stocks are overfished (NMFS 2007).

The striped bass spawning grounds in Albemarle Sound and Chesapeake Bay are critical to the survival of this species and its commercially important fishery. After a collapse of the striped bass fishery in the late 1980s, extremely restrictive harvest regulations have resulted in a surprisingly strong and rapid comeback of the species and reopening of the fishery.

Head of a North Atlantic right whale marked by the characteristic whitish bump-like features called callosities.   *Photo:* Sam Fried/Photoresearchers

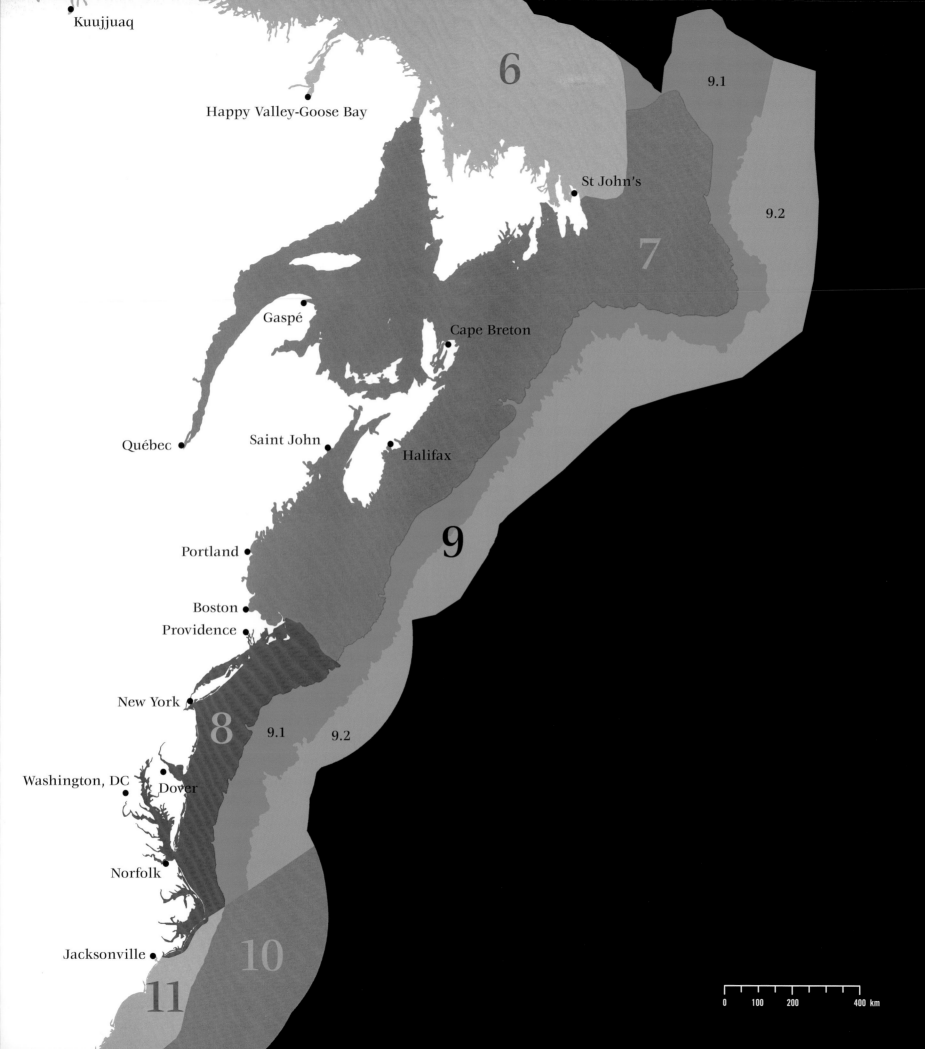

Kuujjuaq

Happy Valley-Goose Bay

6

9.1

9.2

St John's

7

Gaspé

Cape Breton

Québec

Saint John

Halifax

9

Portland

Boston

Providence

New York

8

9.1

9.2

Washington, DC

Dover

Norfolk

Jacksonville

10

11

| 0 | 100 | 200 | | 400 km |

9. Northern Gulf Stream Transition

Level II seafloor geomorphological regions include:

9.1   Northern Gulf Stream Transition Slope
9.2   Northern Gulf Stream Transition Plain

No Level III coastal regions are found in this region.

Regional Overview

The Northern Gulf Stream Transition Region—a region of open ocean in North Atlantic lying adjacent to and affected by the Gulf Stream[10] to the east, the Labrador Current to the north and west, and the Acadian and Virginian Atlantic Regions (the neritic regions) to the west—is an important and highly productive region for over 250 fish species, including bluefin tuna and Atlantic white marlin. Extending from waters off Cape Hatteras to north of the Grand Banks, the region does not border any continental landmass, but overlies several important bathymetric features of the northwestern Atlantic, such as the Canyon Lands, Pickett Seamount and the New England Seamount chain. This once remote marine region is becoming increasing under pressure from fishing due to new deep-sea and open ocean technologies and equipment.

Physical and Oceanographic Setting

The Northern Gulf Stream Transition Region, lying east of regions 7 and 8, begins at the shelf break and extends seaward over North Atlantic deep waters. The region is composed of Gulf Stream-influenced Atlantic water and steep bathymetry, extending from 200 m at the shelf break to nearly 4,500 m at the abyssal plain. At the western edge of the region is the locus of the Canyon Lands—a series of steep canyons, extending from the top of the shelf break through the slope to the abyssal plain. Though the region is not within the Gulf Stream current (the Gulf Stream is south and east of the region), it is strongly influenced by it, and is prone to anti-cyclonic warm core eddy incursions, which spin off of the Stream and meander westward to the shelf break. These warm core rings can be as large as the state of Massachusetts and carry parcels of warm water far northward of their general provenance. Sea surface temperatures of the region can range in January from 10°C at the cold western edge to 18°C at the eastern boundary with the Gulf Stream. In summer, surface temperatures range between 12° and 25°C. Sediment types of the region are mostly silts and clays delivered from the continental land mass.

Biological Setting

Over 250 fish species have been recorded in this important and highly productive area. The deep waters of this region are home to high densities of bluefin tuna and blue and white marlins. Great northern tilefish congregate on the continental slope to

---

10 The actual Gulf Stream and its associated rings extend to the east, beyond the EEZ. Pelagic features such as these meander and move, and are not fixed in time and space.

Fact Sheet

Rationale: characterized by currents and sea surface temperatures from the adjacent Gulf Stream, including moderated water temperatures and the frequent presence of warm core and cold core rings; a pelagic area offshore from the NW Atlantic, extending from the shelf break to the EEZ and northward from Cape Hatteras to where the Gulf Stream diverges to the northeast

Surface: 796,365 km²

Sea surface temperature: 10–18°C (winter), 12–25°C (summer)

Major currents and gyres: warm core rings formed from the adjacent Gulf Stream

Physiography: one of the few non-coastal regions, this ecoregion extends from shelf break to the deep ocean

Depth—shelf (roughly 0–200 m): 0%; slope (roughly 200–2,500/3,000 m): 45%; abyssal plain (roughly 3,000+ m): 55%

Major community types and subtypes: deep ocean benthos, pelagic fisheries, deepwater gorgonian corals, octocoral gardens

Productivity: moderately high (150–300 g C/m²/yr)

Species at risk: sperm, fin, humpback and North Atlantic right whales; loggerhead and leatherback turtles; Atlantic white marlin

Human activities and impacts: Overfishing has affected some species. For instance, the Atlantic white marlin, found throughout the western Atlantic usually above the thermocline in deep pelagic waters, is a victim of overfishing and current stocks are 5–15 percent of carrying capacity. Bluefin and bigeye tuna have also been heavily overfished in the region.

depths of 500 m, and Northern Atlantic right, fin, humpback and sperm whales migrate throughout the region. Several species of sea turtle, including the loggerhead, and leatherback, as well as common, Risso's, bottlenose and Atlantic white-sided dolphin are also commonly observed in this region. The region also represents one of the major breeding areas for the gray seal in eastern North America. Atlantic herring, cod, mackerel, hake and pollock are abundant. These offshore waters are important areas for several avian species, including northern fulmars, storm-petrels, alcids, shearwaters and gannets.

Because the region is a transition zone between the warm tropical waters to the south and the cold Labrador Current to the north, there is often a large migratory component to the species composition of the fish communities. Many of the residents are eurythermic tropical or warm-temperate forms migrating from the south for periods when the water temperature is high. Thus, while species diversity is high, relatively few species are endemic to the area.

Human Activities and Impacts

The Northern Gulf Steam Transition Region encompasses a vast distance, much of which is threatened by overfishing. Where much of this region was not greatly exploited by fishing in the past, new technologies, such as remotely operated vehicles, video, sea surface temperature images, bottom landers, submarines and sonar, as well as sturdier winches, stronger cable, and more powerful engines have allowed fishing trawlers to extend their reach to depths of 900 m and beyond. Moreover, because: a) fishing effort is more difficult to control on the high seas and b) deepwater species are slower growing, longer lived and reproduce later in life, deepwater fisheries are not particularly sustainable. While proposals are being developed to ensure the protection of habitat and sustainability of pelagic fisheries, it is generally accepted that the open water areas are underrepresented and unprotected in much of the northwest Atlantic. Bycatch is also an issue for the Northern Gulf Stream Transition Region. As within the Gulf of Mexico, incidental take and entanglement of sea turtles and other species occurs in this area. Moreover, US Navy sonar exercises also likely affect many marine species (NRC 2003).

This area is also affected by pollution—particularly plastics and hydrocarbons that accumulate in driftlines and at convergence zones. Fish and crustacean eggs and juveniles, as well as juvenile sea turtles and even adult seabirds are affected.

The famous blunt-nosed sperm whales favor the deeper and more temperate oceanic waters.
Photo: Brandon D. Cole

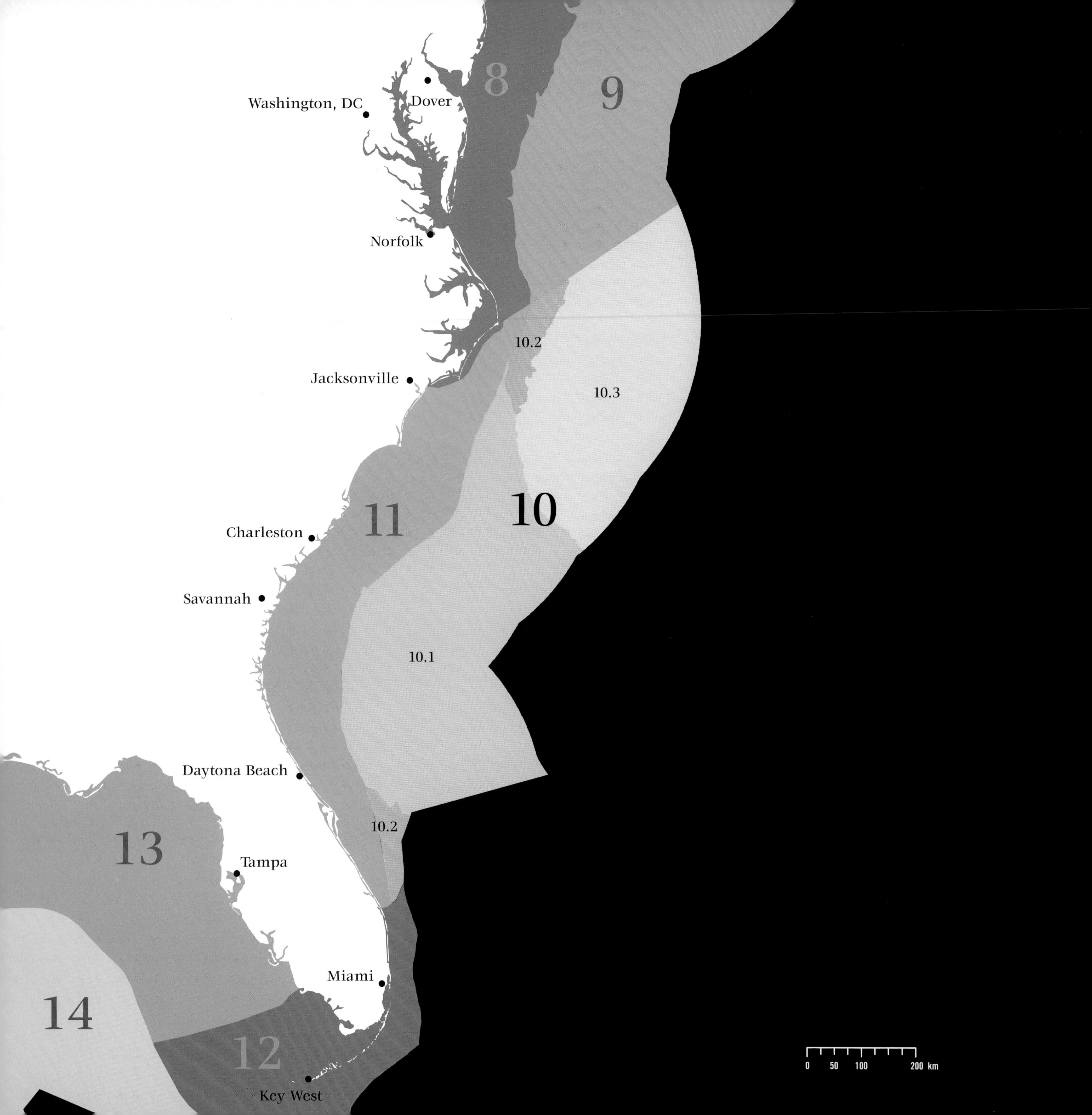

8
9
Washington, DC
Dover
Norfolk
10.2
Jacksonville
10.3
11
10
Charleston
Savannah
10.1
Daytona Beach
10.2
13
Tampa
Miami
14
12
Key West
0 50 100 200 km

10. Gulf Stream

Level II seafloor geomorphological regions include:

10.1   Blake Plateau
10.2   Gulf Stream Slope
10.3   Gulf Stream Plain

No Level III coastal regions are found in this region.

Regional Overview

The Gulf Stream Region is defined and dominated by the Gulf Stream current—a river within the ocean. This offshore region sees many migrating species such as the humpback whale and bluefin tuna, and is home to the only known North American population of wreckfish. Due to the interaction of the Gulf Stream with bathymetric features such as the Charleston Bump, upwelling occurs, enriching surface waters downstream of the Bump. The region starts from the Straits of Florida at its southern extreme, and continues northward and seaward of the coastal Atlantic Bight, following the Gulf Stream current to the Outer Banks of North Carolina and Cape Hatteras, where the region terminates as the current veers northeastward (out of the region of study). The region includes sections of the slope, abyssal plain and several ecologically important bottom features.

Physical and Oceanographic Setting

Region 10 represents the Gulf Stream flow along the southeast US coast—the point where the Stream approaches continental North America most closely. The Gulf Stream forms the western boundary current of the North Atlantic subtropical gyre, flowing offshore up to 2,000 m deep, and following along the edge of the continental shelf. This current, often described as a river within the ocean, is up to 320 km wide, with a core about 144 km in width, and carries warm tropical waters poleward at velocities of up to 2.5 m per second. The current forms from the Florida Current, emanating from the Gulf of Mexico, and forms a jet between the continent and Cuba and the Bahamas that boosts the current's speed along the Florida coast. Along the western wall of the current, eddies often form and spin off in a northerly direction to form warm core rings that enter the adjacent Northern Gulf Stream Transition Region. These oceanographic features, which can be many kilometers in diameter, carry semi-tropical fauna into the cold reaches influenced by the Labrador Current.

The bathymetry of the region incorporates two distinct areas. In the northern section, beyond the wide continental shelf off Virginia and North Carolina, the region's sea bottom slopes steeply to a deep abyssal plain. In the southern section, the shelf of the adjacent region leads to a sharp but relatively shallow drop (200–1,000 m) to the immense Blake Plateau—an area that is about three times the areas of the adjacent shelf. East of the Blake Plateau, the slope drops off steeply to the abyssal plain at

Fact Sheet

Rationale: defined by the Gulf Stream, a large, predominantly northeastward flowing current

Surface: 307,813 km²

Sea surface temperature: avg. 23°C (winter), 27–30°C (summer)

Major currents and gyres: Gulf Stream, upwelling along shelf break

Other oceanographic features: the Gulf Stream, an important, large current flowing from the Florida Straits to the northeast at speeds of 0.15–2.5 m/s, carries warm tropical waters and transports biota far to the north of normal distribution range

Physiography: from shelf break to the deep ocean, including Blake Plateau. The slope below the Gulf Stream is structurally complex with many seamounts, canyons and ridges.

Depth: shelf (roughly 0–200 m): 0%; slope (roughly 200–2,500/3,000 m): 59% including the Blake Plateau; abyssal plain (roughly 3,000+ m): 41%

Substrate type: deep silt and basalt seamounts

Major community types and subtypes: deep ocean benthos, open water nekton communities; benthic community types including deep-sea *Lophelia* coral banks

Productivity: Moderately high (150–300 g C/m²/yr); upwelling along the Gulf Stream front and intrusions from the Gulf Stream cause short-lived plankton blooms.

Species at risk: fin, North Atlantic right and sperm whales; leatherback sea turtle; silver hake; Atlantic white marlin; tilefish; and *Oculina varicosa* coral

Human activities and impacts: Overfishing is a major threat—current stocks of the white marlin are 5–15 percent of carrying capacity.

The interaction of the current with the shelf edge, and with bathymetric features such as the Charleston Bump at the South Carolina/Georgia border, causes upwellings to occur, enriching the surface waters downstream (north) of the feature with nutrient rich deeper water. Consequently, the waters around the Bump are very productive, with a complex food web supporting one of the most popular fishing areas of the South Atlantic Bight (in region 11, adjacent). The eddies generated by the motion of the current past such bathymetric features are an important means of transporting reef fish northward to shelf habitats off North and South Carolina.

Below the Gulf Stream, an extensive system of deepwater coral banks, dominated by *Lophelia* and *Dendrophelia,* form reefs on and below the shelf break. In particular, stony corals (*Lophelia pertusa* and *Enallopsammia spp.*) are found in the Straits of Florida and on the Blake Plateau off North Carolina, South Carolina, Georgia and Florida. These structures support a high diversity of fish and macro-invertebrate species, and appear to be the most extensive reef formations known from the northwest Atlantic.

a depth of approximately 5,000 m. Sediments are composed of silty clays. Sea surface temperatures range from 23°C in winter to 27–30°C in summer.

Biological Setting

The biology of this region is similar to, and in many ways a continuation of, region 9 to the north. The deep waters support high densities of bluefin tuna, Atlantic blue and white marlins, and tilefishes on the continental slope. The Charleston Bump/ Blake Plateau area is also home to the only known North American population of wreckfish—a slow-growing, late-age maturing and long-lived species that prefers deep reef habitats. In fact, this area is the only documented spawning area for the fish in the North Atlantic. The North Atlantic right, fin, humpback and sperm whales as well as the leatherback sea turtle migrate throughout the region, its southern section representing the southern limit of many of these whale species. Because the Gulf Stream current governs much of the biology of the region, and also because the current is maintained offshore by the shallow continental shelf barrier, the biology of this region is very different from that of the adjacent Carolinian Atlantic Region. The Gulf Stream carries water of moderate temperature far north of this region's limit, extending the range for many subtropical species such as dolphinfish, wahoo, butterflyfish, wrasses, and goatfishes.

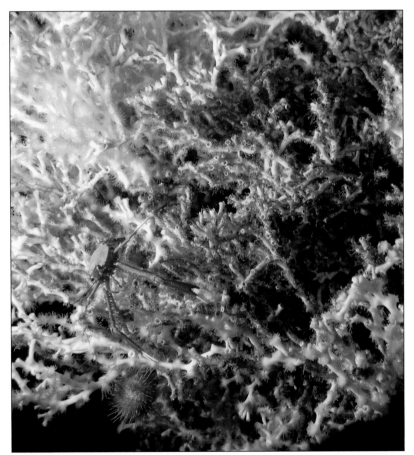

Among the numerous organisms that take refuge in deep-sea lophelia coral reefs are squat lobsters (*Eumunida picta*), only recently documented in the Gulf of Mexico. *Photo:* Ken Sulak, USGS (*Life on the Edge 2004 mission*)

A number of commercially and recreationally important fisheries occur along the edge of the Gulf Stream, where the current interacts with the shelf break creating upwelled hotspots of high biological productivity. Tuna (bluefin, bigeye) is caught in abundance in this region, as is the white marlin, Atlantic croaker and various species of tilefish. There is concern over overfishing (swordfish, in particular) and bycatch of untargeted species (such as undersized swordfish, marlin and sailfish) in the region around the Charleston Bump, in particular. Recreational and commercial fishing, as well as commercial shipping carries many vessels through the region, creating the potential for severe damage to the ecosystems. For example, fishing activities have destroyed the *Oculina* or ivory tree coral reefs off the north coast of Florida. Ivory tree coral, a very slow-growing deepwater coral, is fragile and very difficult to re-establish once damaged. Fishing of the reef fish—grouper, wreck-fish—and for migratory pelagic species has decimated large parts of the reef over the years. To aid in recovery of this unique and ecologically important habitat, a large portion of the area was closed to fishing in 1994 as the Oculina Bank Habitat of Particular Concern, the first deepwater site to be granted such protection in the eastern United States. The reef is currently protected in the shallowest part of its range, but much of the remaining reef exists at or beyond the shelf break (and overfishing of long-lived groupers still occurs).

The leatherback turtle, the largest of all turtles, is seen here off the Florida coast with freeloading sucker fish (remoras) attached. *Photo:* Michael P. O'Neill

8

Jacksonville •

11.1.2

11.1.1

11.1

11.2

Charleston •

Savannah •

10

11

11.1.1

Daytona Beach •   11.1

11.1.2

Tampa •

13

Miami •

0 25 50   100 km

11. Carolinian Atlantic

Level II seafloor geomorphological regions include:

Level III coastal regions include:

Regional Overview

The Carolinian Atlantic Region, with its numerous wetlands and tidal marshes, serves as a nursery area for many marine fishes and is an important region for shellfish. It is also highly stressed by land-based activities such as factory farming. In addition, the region combines coastal development and severe hurricanes with the most disagreeable results. The region is bounded to the south by the Atlantic coast of Florida offshore of Palm Beach, where the shelf widens and the Gulf Stream begins to diverge from the coast, and to the north by the Outer Banks and Cape Hatteras. The region is defined by a broad continental shelf that extends up to 150 km from the coast at Georgia and by several Coastal Plain watersheds that terminate at the coastal margin. The Carolinian Atlantic Region is bounded to the east by the edge of the Florida-Hatteras slope and the Gulf Stream.

Physical and Oceanographic Setting

The Carolinian Atlantic Region is characterized by a wide shallow continental platform, with a 20-meter isobath extending 25 km from shore, and a shelf break at a depth of 60 m occurring nearly 100 km offshore. The area is fed by fresh water discharged from several coastal plain estuaries along the North and South Carolina and Georgia coasts. The most important among these rivers are, in South Carolina, the Pee Dee River feeding Winyah Bay and the Cooper-Santee Rivers that partially supply Charleston Harbor, and the Savannah and Altamaha Rivers in Georgia. The Florida coast is noteworthy for possessing only a single major river in this region—the St. Johns River. Moderate oceanic currents tend to flow parallel to the coast, generally northward year-round in the upper half of the region, and generally southward in the lower half. Confined to the area seaward of the shelf break, the Gulf Stream flow does not influence the current structure of this region. The Carolina shelf water generally is characterized by salinities greater than 35 PSU. Tides can be significant, ranging from 1 to 3 m, with maximum tidal amplitudes occurring along the South Carolina and Georgia coasts. Sea surface temperatures range from 15 to 22°C in winter and average about 28°C in summer. Geologically, almost the entire coastal region is composed of non-resistant rocks, with little relief along the coastal margin. Barrier islands occur throughout the region, but notably in North Carolina at the Outer Banks, and at Cape Canaveral in Florida. Sediments are chiefly alluvial deposits of sand and silty sand.

Fact Sheet

Rationale: region defined by similar sea surface temperature, faunal composition and oceanographic currents. The northern boundary represents a major biogeographic transition.

Surface: 125,606 km²

Sea surface temperature: 15°–22°C (winter), 28°C (summer)

Major currents and gyres: nearshore estuarine freshwater inflow; weak southerly longshore currents.

Depth: shelf (roughly 0–200 m): 92%; slope (roughly 200–2,500/3,000 m): 8%; abyssal plain (roughly 3,000+ m): 0%

Substrate type: non-resistant rock, coastal plain sands and depositional silt-clays

Major community types and subtypes: large estuaries, coastal salt marshes, coastal embayments, barrier islands, river mouths, sandy beach, soft bottom, oyster reefs, tidal channels, seagrass beds

Productivity: moderately high (150–300 g C/m²/yr)

Species at risk: North Atlantic right and fin whales; West Indian manatee; green, hawksbill, leatherback and loggerhead sea turtles; diamondback terrapin; shortnose and Atlantic sturgeons; speckled hind; barndoor skate; night, dusky and sand tiger sharks; and the *Oculina varicosa* coral

Important introduced and invasive species: red lionfish

Human activities and impacts: fishing, tourism, commercial shipping, navigation and agriculture. Pig farming in the Carolinas has created highly contaminated surface runoff and eutrophic conditions in rivers and estuaries.

Biological Setting

Owing to the flat coastal plain, gentle topography, and numerous rivers flowing from the Appalachians to the east, this region is characterized by forested and non-forested wetlands and tidal marshes. The amplified tidal range and twice-daily inundation of the flat coast also contributes to the formation of the large areal extent of tidal wetlands and flats. This physical setting strongly affects the biology of the region, with high concentrations of salt marshes, numerous nursery areas for estuarine-dependent species, and many marine fishes that utilize the coastal zone of the region. Eastern oyster reefs form at the mouths of tidal creeks flushed by the strong tidal subsidy and white, brown, and pink shrimp occur in large concentrations in the Florida-Georgia Bight from the coastline to the shelf edge. Blue crab and spiny lobsters occur in non-commercial concentrations offshore. The Atlantic sturgeon uses virtually every one of the Carolinian Atlantic Region estuaries to spawn. American shad, Atlantic menhaden, bluefish, black sea bass and bigeye scad are resident or migrant species that heavily utilize this area. There are resident river populations of striped bass in each of major estuaries of the Piedmont. The offshore waters are moderately productive, with short-lived plankton blooms associated with upwelling along the Gulf Stream front.

The Gray's Reef National Marine Sanctuary is located just off the coast of Georgia in about 18 m of water. This area contains the largest sandstone reefs in the southeastern United States, consisting of sandstone ledges up to about 3 m high, with sandy flat-bottom troughs between them. The region is extremely diverse, biologically, and attracts sportfishing and diving. Near the continental shelf break, off the Florida coast, are the world's only known ivory tree coral (*Oculina*) banks spanning this region and the adjacent Gulf Stream Region.

Human Activities and Impacts

The Carolinean Atlantic Region supports important urban areas, busy shipping ports, and important commercial and recreational fisheries resources. The coastal population has increased in the region by 160 percent between 1960 and 2000. Development along the coast has introduced a number of anthropogenic stresses to the region—including to both its nearshore and offshore ecology. Examples include elevated nutrient loads associated with urban waste and agricultural activities, pesticide inputs, and development of infrastructure on barrier island systems. In the Carolinas, particularly in North Carolina, the recent consolidation of small pig farms into large factory farms of 2,000 hogs or more has led to environmental problems in the coastal zone (Shaw 2000). The 10 million hogs in North Carolina are concentrated in the eastern part of the state near sensitive wetlands and watersheds. The

Prehistoric-looking horseshoe crabs laying eggs at high tide in Delaware Bay.   *Photo:* Stephen J. Krasemann/DRKPhoto

The principal threat faced by the hawksbill turtle is being hunted for its valuable shell. *Photo:* Claudio Contreras

high density of hogs requires the storage of hog waste in large pits that are prone to seepage and overflow into waterways, especially when heavy storms strike. High nitrogen and phosphorus content readily fertilizes the ocean waters, promoting phytoplankton growth, harmful algal blooms, reduced water clarity and shifts to undesirable species compositions. Fish lesions become common during and after storms and high flow events. There is some evidence that eutrophic conditions are promoting blooms of the toxic form of a dinoflagellate called *Pfiesteria piscicida*, which has been responsible for fish disease events, fish kills and sickness in humans. The destruction of wetlands on the shores has magnified these effects. Overall, the ecological condition of estuaries in the region is fair to good (EPA 2005). **Forty percent of the estuarine area fully supports human and aquatic life uses, 37 percent is threatened for human and aquatic life use, and 23 percent is impaired for these uses (EPA 2005). Four out of 21 major fish stocks are overfished and seven are of unknown status (NMFS 2007).**

The region is subjected to catastrophic natural events such as severe hurricanes. Housing and road development in the barrier dune systems have resulted in beach and barrier island degradation, making the region more susceptible to washout from hurricanes and habitat loss. Development in flood plains and wetland areas can disrupt sediment nourishment of marshes and lead to wetland loss. The result is conversion of coastal marshes to open water and loss of the nursery function provided by marsh habitat. Hurricanes and strong storms can rip large areas of deteriorating marsh from the land, creating coastal zones devoid of important marshes. Hurricanes also exacerbate overflow from manure lagoons and the runoff of farm fertilizers into waterways. Hurricane Floyd struck North Carolina in September 1999 and caused massive overflows into the Neuse River. Due to the extensive presence of barrier islands along the coasts in the region, riverine outflow and the accompanying contaminants are effectively trapped in the back-island lagoons between the islands and the mainland, held close to shore, where flushing by ocean circulation is slowed.

Channelization, dredging, and dumping also take their toll on the region. Owing to the fluvial and highly sedimentary nature of the coast around navigation routes, dredging is an ongoing activity, which impacts natural communities. Dredged material is disposed of in dozens of sites close to shore throughout the region. Offshore, ocean dumpsites are found throughout the region, as well as large munitions and explosives deposits on the slope off the Georgia-South Carolina coasts that have resulted from the numerous military bases and installations in the coastal zone.

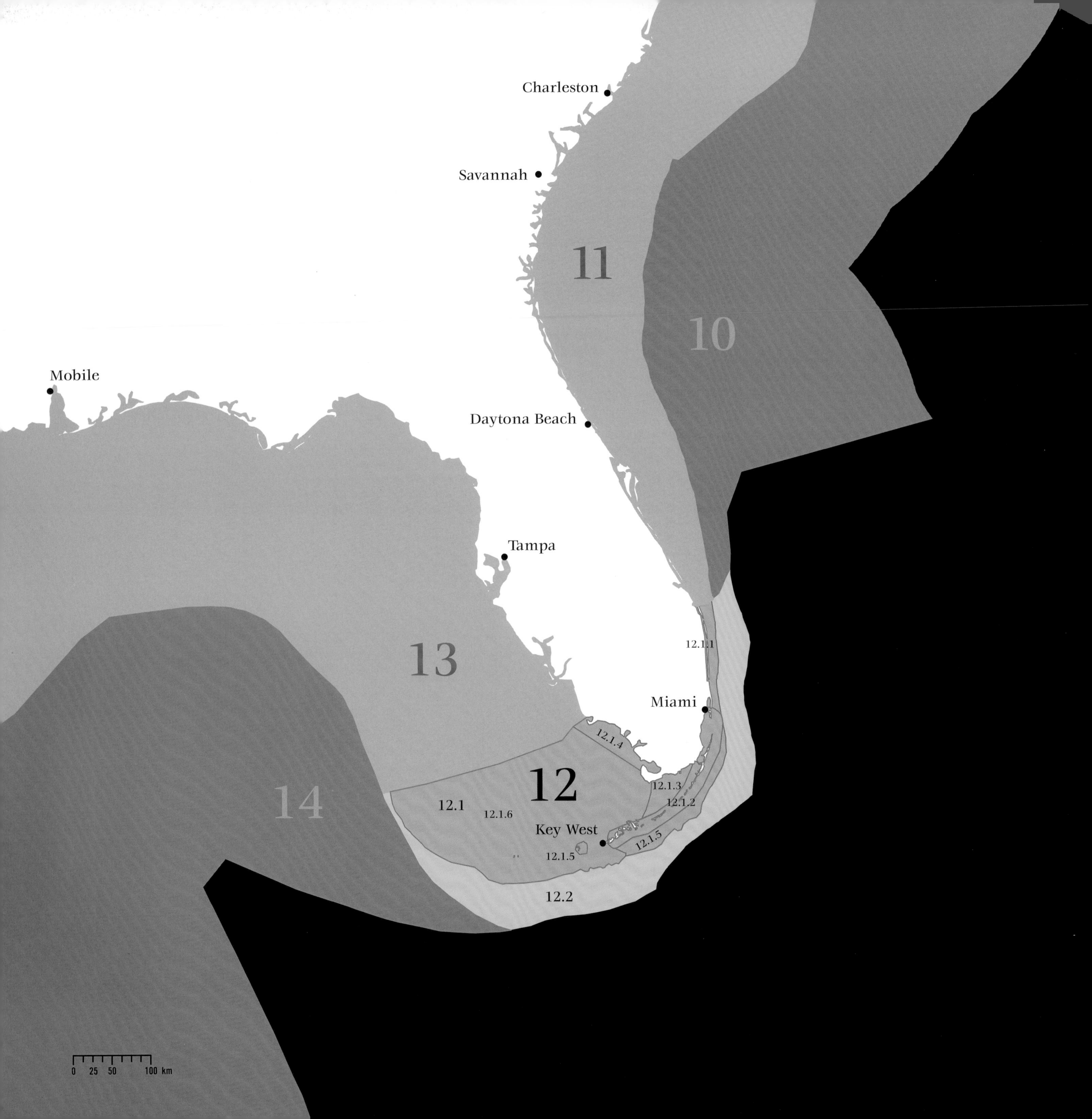

Charleston

Savannah

11

10

Mobile

Daytona Beach

Tampa

13

12.1.1

Miami

14

12

12.1.4

12.1

12.1.6

12.1.3

12.1.2

Key West

12.1.5

12.1.5

0 25 50   100 km

12.2

12. South Florida/Bahamian Atlantic

Level II seafloor geomorphological regions include:

12.1   South Florida/Bahamian Shelf
12.2   South Florida/Bahamian Slope

Level III coastal regions include:

12.1.1   Southeast Floridian Neritic
12.1.2   Florida Keys
12.1.3   Florida Bay
12.1.4   Shark River Estuarine Area
12.1.5   Dry Tortugas/Florida Keys Reef Tract
12.1.6   Southwest Floridian Neritic

Regional Overview

With mangrove forests, sandy beaches, seagrass beds, and coral reefs, the South Florida/Bahamian Atlantic Region is a very complex and important region for tropical species, but also one much stressed by anthropogenic forces. The region is small, with generally clear waters, coral reef formations, carbonate substrate, and is tropical to subtropical in character. Climate, substrate and biota are influenced primarily by the Gulf Stream. The region includes coastal waters off southern Florida, the nearshore region where the continental shelf break and the Gulf Stream most closely approach the coast. This ecoregion extends to Naples, Florida, on the Gulf of Mexico coast, although some habitat types may extend as far north as the Anclote Keys, the northern extent of mangrove habitats.

Physical and Oceanographic Setting

Oceanographically, this region is very complex, with physiography and hydrology converging in a unique and biotically rich area. Physically and biogeographically, the area in south Florida is similar to the Bahamas, although freshwater runoff is much lower in the archipelago. In extent, the continental shelf ranges from an extreme of 150 km wide off the west coast of Florida to 5 km wide off the east coast. In between, there are steep depth gradients just off the reef tract at Pourtales Terrace. Florida Bay is a carbonate-based shallow estuary receiving fresh water input from the Everglades, whereas Biscayne Bay on the east coast receives water from canal-delivered runoff. The Gulf Stream moves through the region from west to east through the Straits of Florida, and turns north into the Atlantic. The Loop Current, meso-scale gyres in the Keys, and Hawk Channel in the Keys, all contribute to the complex physiography of the region and the ability of the area to retain marine larvae and propagules. Hurricanes are a significant force in the ecological dynamics of the area, with a probability of occurrence of 8–16 percent across the region. Their effects include massive fresh water input, resuspension and scouring of the ocean bottom, extirpation of the seagrass communities and destruction of human infrastructure.

Biological Setting

Within Florida Bay, in the interior of the Florida Keys, the region is shallow, micro-tidal and estuarine in character. The extensive seagrass communities in the Bay are the largest in the northern hemisphere; turtlegrass, shoalweed, manateegrass, and in the areas of generally lower salinity, widgeongrass carpet much of

Thanks to conservation efforts, there is a resurgence of reef fish and corals in the Dry Tortugas National Park, Florida. *Photo:* NOAA

Fact Sheet

Rationale: defined by subtropical climate and sea surface temperature, warm water fauna

Surface: 82,426 km²

Sea surface temperature: 22.5°C (winter), 28°C (summer)

Major currents and gyres: Gulf Stream

Other oceanographic features: frequent fronts and hurricanes

Depth: shelf (roughly 0–200 m): 64%; slope (roughly 200–2,500/3,000 m): 36%; abyssal plain (roughly 3,000+ m): 0%

Major community types and subtypes: mangrove forests, seagrass beds, coral reefs, sand banks, as well as a matrix of habitats of soft and consolidated bottom types, caves and crevices. West Palm Beach is the northernmost extent of *Acropora* corals.

Productivity: moderately high (150–300 g C/m²/yr)

Endemics: The saltmarsh topminnow is an endemic species found as far south as the Perdido, Escambia and East Bays of Florida in salt, brackish and freshwater marshes.

Species at risk: West Indian manatee; bald eagle; piping plover; Cape Sable seaside sparrow; roseate tern; American crocodile; American alligator; loggerhead, Kemp's ridley, hawksbill and green sea turtles; pink shrimp; corals and sponges; smalltooth sawfish, speckled hind, opossum pipefish, dusky shark, sand tiger shark, night shark, elkhorn and staghorn coral, Johnson's seagrass

Important introduced and invasive species: In the wetlands environment, invasives include purple loostrife, curly-leaved pondweed, and torpedograss and the water hyacinth. At least four non-native fish species are established in estuaries and 15 species of reef fishes have been observed, including the red lionfish, which has become established.

Human activities and impacts: Over-development of shore and dune systems and destruction of dune vegetation has accelerated erosion and overwash and increased nutrient inputs. Channelization of the Everglades cordgrass wetlands has altered the salinity and ecology of Florida Bay. Destruction of nesting habitat is a major threat to sea turtles. White marlin, found throughout the western Atlantic usually above the thermocline in deep pelagic waters, is a victim of overfishing—current stocks are 5–15 percent of carrying capacity; overfishing, shipping lanes prone to bilge and waste discharges also threaten numerous pelagic and demersal species in the region; destruction of marsh habitat by conversion or land erosion is eliminating the habitat of the saltmarsh topminnow.

the sandy to silty bottom. The coastal margin is fringed by mangroves, including red, black, white and button species. Some coral outcroppings are evident.

The region's reef tract is a semi-continuous reef, the third-longest in the world, located on the seaward side of the Florida Keys and extending to the outer edge of the narrow continental shelf. The reef continues as far north as the Ft. Lauderdale area, which is the northern limit of *Acropora* coral species. Hard bottom communities, calcareous carbonate sands, and coral reef development typify the area of the reef tract and seaward. Mangrove forests populate the coastal zone of this subregion and the water column has oceanic salinities and tidal amplitudes reaching 1 meter. Caribbean spiny lobster and blue crab occur throughout the region to the shelf break. Pink shrimp occur in large concentrations to the shelf break and royal red shrimp to a depth of 500 m on the oceanside. Mackerel, Atlantic manta, bluefish, angelfish, amberjack, bonefish and snapper occur throughout the

region, with the latter two heavily concentrated in the nursery areas of Florida Bay. The Atlantic and Gulf coasts are particularly important habitat for sea turtles (loggerhead and green) and Florida Bay and southern Everglades is an important breeding area for crocodilians (American alligator and American crocodile).

Human Activities and Impacts

Often described as intensely used and managed, this region borders the four most densely populated counties in Florida. Coral reefs provide a foundation for fisheries and tourism-based economy that generated an estimated 71,000 jobs and US$6 billion in economic activity in 2001 for southeast Florida and the Florida Keys (Johns *et al.* 2001). The region is under anthropogenically induced stresses such coastal development,[11] septic system-based coastal water contamination, runoff, reductions in freshwater inputs, pollution, damage from ship groundings and anchor scars, and unsustainable fishing practices. Moreover, the same coral bleaching and mortality observed in the Caribbean Sea ecosystems is occurring in the shallow waters of the South Florida/Bahamian Region, with events in the region having increased during the 1980s and 1990s. In some areas, nutrient enrichment is causing reefs to be overgrown by macroalgae—with devastating effects on the systems the reefs support. Mass seagrass mortality in Florida Bay has also resulted in a cascade of effects that include biomass decomposition and nutrient release, increased sediment resuspension and reduction in water clarity, phytoplankton blooms, and reduced fisheries. The region is further stressed by reductions in fresh water input to Florida Bay due to channelization, overdevelopment, agriculture and urbanization of the watershed. This has resulted in hypersalinity events, encroachment of seawater into the estuary and the aquifer, and species replacement. Currently, the Comprehensive Everglades Restoration Plan is being developed to try and reverse some of these deleterious impacts.

---

11 Habitat loss occurs from development such as coastal construction, building of seawalls, jetties, and causeways, dredging, infilling of wetlands, eutrophication from overburdened septic systems.

Florida manatee in the Blue Springs State Park, Florida. Habitat destruction and collisions with boats kill most manatees in the wild before they have reached 30 years of age. *Photo:* Doug Perrine/DRK PHOTO

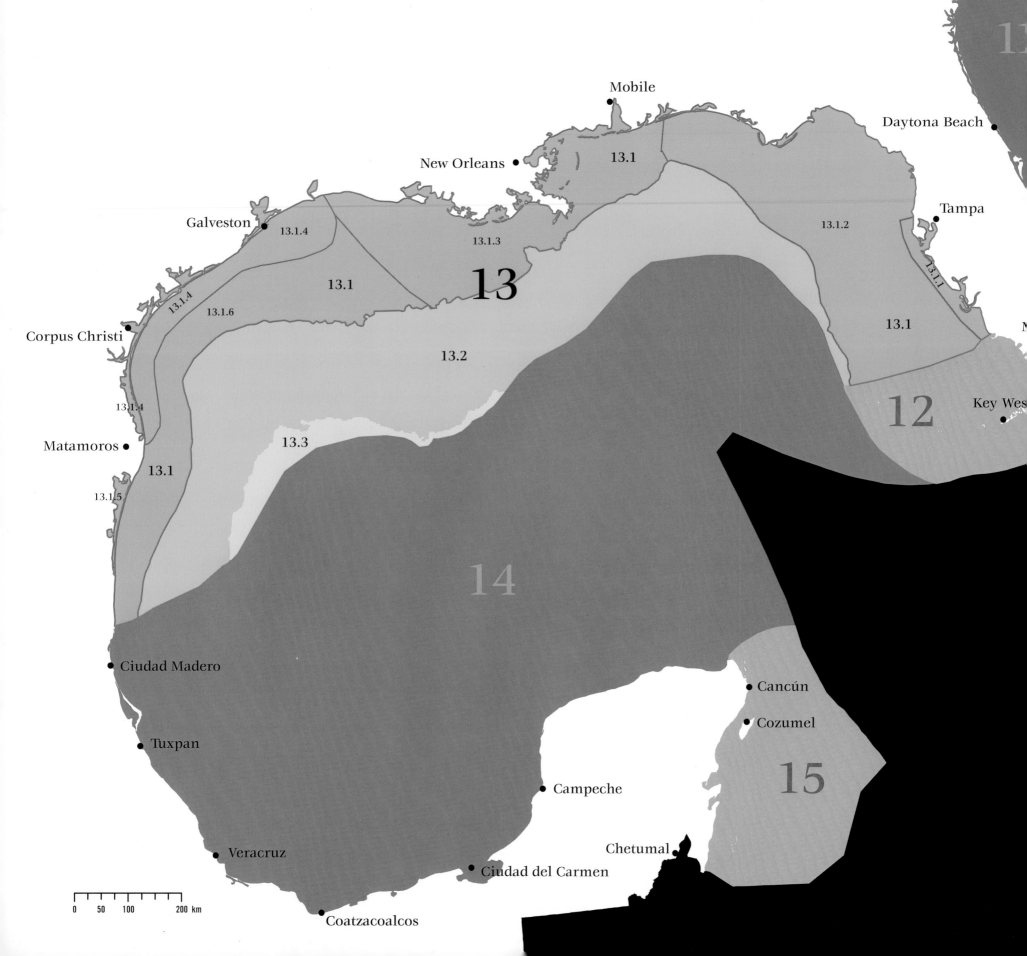

13. Northern Gulf of Mexico

Level II seafloor geomorphological regions include:

13.1   Northern Gulf of Mexico Shelf
13.2   Northern Gulf of Mexico Slope
13.3   Gulf of Mexico Basin

Level III coastal regions include:

13.1.1   Western Florida Estuarine Area
13.1.2   Eastern Gulf Neritic
13.1.3   Mississippi Estuarine Area
13.1.4   Texas Estuarine Area
13.1.5   Laguna Madre Estuarine Area
13.1.6   Western Gulf Neritic

Regional Overview

The northern Gulf of Mexico Region (hereafter referred to as the northern Gulf), contains over 60 percent of the tidal marshes of the United States, freshwater inputs from thirty-seven major rivers, numerous nursery habitats for fish, the Flower Gardens Banks, and also the so-called 'dead zone' resulting from increased loads of organic material from the extensive Mississippi River watershed. The region extends from Gullivan Bay on the west coast of Florida to north of the Rio Pánuco in the state of Tamaulipas, Mexico, and includes the coastal areas of the US states of Alabama, Mississippi, Louisiana, and Texas. The region comprises the northern portion of the Gulf of Mexico (hereafter referred to as the greater Gulf)—a semi-enclosed sea with tropical currents and a high nutrient load. Most of the oceanic input to the greater Gulf is from the Caribbean Sea through the Yucatan Channel, forming the Loop Current that winds north then east through the greater Gulf, outflowing through the Straits of Florida. A broad continental shelf covers about a third of the greater Gulf.

Physical and Oceanographic Setting

The northern Gulf is characterized as semi-tropical due to the seasonal pattern of its temperature regime, which is influenced mainly by tropical currents in the summer and temperature continental influences during the winter. It has a distinct sea surface temperature gradient from north to south (up to 7°C) in winter. The Gulf of Mexico presents a seasonally changeable wind regime with predominance of the northeast trades.

The northern Gulf is characterized physiographically by a broad continental shelf extending out to 250 km from the coastline, a steep continental slope and a small section of the large central abyssal plain of the greater Gulf. Distinctive bathymetric and morphological processes and features have had a great influence on the makeup and functioning of the region. For example, the Flower Gardens Banks—surface expressions of salt domes off the Louisiana/Texas coast—are home to the northernmost coral reefs in the United States, have natural gas seeps, but also numerous tropical fishes, manta rays, turtles and sharks. The entire southern portion of the state of Louisiana and part of eastern Texas was formed by delta building and river-switching processes of the Mississippi River. The Chenier Plain and Mississippi River Birdsfoot Delta were deposited by deltaic and plume transport processes of the Mississippi River. And finally, the extensive barrier island system that runs from the Florida Panhandle through Alabama,

Fact Sheet

Rationale: The region is defined physiographically by the enclosed nature of the Gulf of Mexico as well as by a sea surface temperature differential with the southern Gulf of Mexico in the winter.

Surface: 578,294 km²

Sea surface temperature: 14–24°C (winter), 28–30°C (summer)

Major currents and gyres: Loop Current (cyclonic), Florida Current, Tamaulipan Gyre (anticyclonic)

Other oceanographic features: semi-enclosed basin with high nutrient load, tropical currents in summer and temperate influence during the winter; home to warm-temperate biota; rainfall is less than 700 mm/yr

Physiography: broad continental shelf along Florida, narrowing to a steep, thin shelf fronting the Mississippi River outlet, widening again along the Texas coast, narrowing in the Mexican southern portion; most of the nearshore waters are divided into bays and estuaries behind barrier islands

Depth: shelf (roughly 0–200 m): 56%; slope (roughly 200–2,500/3,000 m): 40%; abyssal plain (roughly 3,000+ m): 4%

Substrate type: sand barriers; silt and mud, with clays in central Gulf coast; sandy muds in Texas; and sand and carbonate muds in Florida

Major community types and subtypes: mangrove ecotone, patchy seagrass communities, oyster reefs, isolated salt domes with coral reefs and deep *Lophelia* coral mounds, deltaic systems, coastal lagoons, estuaries, salt marshes, river inlets, dwarf mangroves, serpulid worm reefs

Productivity: moderately high (150–300 g C/m²/yr)

Species at risk: Kemp's ridley, green and loggerhead sea turtles, Gulf sturgeon, speckled hind, dusky shark, sand tiger shark, night shark, largetooth sawfish, smalltooth sawfish, Texas pipefish, opossum pipefish, dwarf seahorse, diamondback terrapin, Alabama shad, and saltmarsh topminnow

Key habitat: areas of the broad continental shelf, coastal lagoons and estuaries, river inlets, cypress swamps, mangroves swamps, seagrass beds, oyster reefs, tidal freshwater grasses, salt marsh, tidal freshwater marsh flats, intertidal scrub forest, muddy bottom habitats, Coquina beach, rock marshes and bars, intertidal/subtidal beaches and bars, serpulid worm reefs

Human activities and impacts: tourism expansion; urban expansion; ports; oil and gas exploration and recovery; liquefied natural gas (LNG) terminals; shipping; coastal industrialization and infrastructure; natural and anthropogenic loss of wetlands; coastal erosion; fishing activities cause additional impacts, including benthic habitat alteration by shrimp trawling; fishing along slope kills sea turtles and other threatened species as bycatch; the goliath grouper was subjected to severe fishing pressure on spawning aggregations; destruction of marsh habitat by conversion or land erosion is eliminating the habitat of the saltmarsh topminnow; ecological changes have also been wrought by the deliberate introduction of whiteleg shrimp

Texas, and northern Mexico—created by longshore transport and deposition of sands—forms many lagoons and sheltered areas that serve as refugia and spawning grounds.

Muddy clay-silts and muddy sands dominate bottom substrates of the region throughout the entire shelf, slope and plain off the Louisiana, Texas and Mexico coasts. From Alabama east to Florida, sand, gravel and shell dominate the region, and on the shelf off Florida, the carbonate limestone substrate is interspersed with gravel-rock and coral reefs. The carbonate limestone substrate of the Western Florida Estuarine Area (Level III region), however, is overlain by sand and silt and supports extensive seagrass beds dominated by turtlegrass that in turn are nursery, feeding and spawning areas for several important fishes.

The entire region features twice-daily tides of variable low amplitude, generally between 5 and 30 cm. Several persistent currents—including the Loop Current, the Coastal Countercurrent, and the Florida Current—mark the greater Gulf waters. The Gulf Stream, which has such an immense effect on the waters of the Atlantic, also originates in this region. In the eastern Gulf (level III region), the eastern edge of the Loop Current interacts with the shallow shelf to create zones of upwelling and onshore currents. These are nutrient-rich events and promote high phytoplankton growth and support high biological activity.

The region in the central north is strongly dominated by the Mississippi River and its tributary, the Atchafalaya River, which carries 30 percent of the flow of the Mississippi. Together these rivers' discharge exceeds all other Gulf rivers by an order of magnitude. In addition to these major rivers, several other large rivers and estuaries are found in the region. The quantity and quality of the fresh water coming from these systems significantly impact the coastal physico-chemical characteristics and biological communities of the region.

Along with several smaller fresh water sources, these rivers deliver the greatest volumes of freshwater to the northern Gulf during the wet/flood season from May through November. During this period, the region is marked by large plumes of turbid water which become entrained in a westward-flowing longshore current, carried westward to the Texas border, and then join the Loop Current and be carried eastward to southern Florida.

Recently, the delivery and deposition of increased loads of terrestrial organic material from the extensive Mississippi River watershed has become a concern in the greater Gulf. These increased loads affect even relatively deep offshore waters, and have resulted in accumulations of organic material, bacterial degradation, and increased oxygen demand, often resulting in severe oxygen depletions in the bottom waters and the frequent appearance of a so-called 'dead zone'—an area where large populations of benthic fauna die because of low oxygen concentrations.

In the western Gulf of Mexico, freshwater input is low and bracketed by the low-flowing Rio Grande (US)/Rio Bravo (Mexico) to the south, the Brazos (US) to the north, and subsidized by Laguna Madre (Mexico/US). The low volume of fresh water input from this system results in clear, high salinity water in this region of low tidal amplitude. The Laguna Madre, in particular, is noted for the long extent of coastline with almost no fresh water input. The region is characterized by a number of estuaries, mostly behind barrier island formations. The micro-tidal western gulf results in a low-energy, broad sandy beach substrate. Seasonally low fresh water input in this highly evaporative area can result in hypersaline conditions and salt pan formation.

A major climatic feature of the greater Gulf and adjacent areas is the occurrence of hurricanes, which greatly affect the physical, biological and human systems of the region. Several severe hurricanes have caused widespread disaster and loss of life along the Gulf coast; however, most biological systems recover relatively quickly from hurricane impacts. The passage of strong wind and storm events are thought to be important to the ecology of this otherwise low-energy region because these episodic inputs of energy rework sediments, redistribute biological seed material and remove accumulated toxics, promoting healthier communities.

Biological Setting

In terms of climate, the region is considered semi-tropical to tropical, and consequently the coastal communities range from salt marshes to seagrasses, and mangrove systems to salt pans, with scarce and isolated coral reef formations. Key habitats in this diverse areas include areas of the broad continental shelf, coastal lagoons and estuaries, river inlets, bald cypress swamps, mangrove swamps, seagrass beds, oyster reefs, tidal freshwater grasses, salt marsh, tidal freshwater marsh flats, intertidal scrub forest, muddy bottom habitats, Coquina beach, rock marshes and bars, intertidal/subtidal beaches and bars, and serpulid worm reefs. Productivity in the northern Gulf of Mexico ranges from eutrophic in coastal waters to oligotrophic in the deeper ocean.

The region is one of the largest estuarine areas in the US, second only to Alaska, and contains over 60 percent of US tidal marshes. The delivery of high loads of terrestrial and agricultural nutrients and organic matter fuels plankton blooms and generally high biological productivity throughout the region. The nutrient input of estuaries support huge commercial fisheries—such as those of spot croaker; Atlantic menhaden; mullet; pink, brown and whiteleg shrimp—and recreational fisheries. Almost all commercially and recreationally important fish and shellfish in the region are estuarine-dependent at some time during their life cycle and use these important areas as spawning, nursery or feeding grounds. The estuarine areas also provide habitat for several threatened and endangered marine mammals, reptiles, fishes and invertebrates.

Along the coastal margin, communities composed of red, black, white and button mangroves thrive in south Florida from Florida Bay to Cape Romano (mainly within South Florida/Bahamian Atlantic region) and in smaller extensions in Texas, and northern Mexico. Scrub mangrove communities occur in Louisiana—the northern extent of mangroves in North America—limited there by temperature.

In many areas of the coastal northern Gulf, particularly in Louisiana, Texas and northern Mexico, brackish and salt marsh vegetation occurs, dominated by needlegrass rush, saltmeadow cordgrass, and smooth/saltmarsh cordgrass. Extensive seagrass beds inhabit much of the shallow coastal margin. Here communities are often dominated by turtlegrass (which harbor many important nekton species), but also include shoalweed, manateegrass, and Johnson's seagrass. In the areas of generally lower salinity, widgeongrass is prevalent. Benthic algae, while less useful as a food source, occurs throughout the entire region from the land margin to the edge of the continental shelf. Blooms also occur in upwelling areas along the Florida shelf break and the Texas shelf. Phytoplankton blooms are prevalent in the areas around all estuarine and freshwater inputs around the Mississippi River and the west coast of Florida.

Aerial view of the Laguna Madre estuary, a unique hypersaline ecosystem. This wetland—the most important in North America—hosts a wide array of biodiversity. *Photo:* Patricio Robles Gil

The Kemp's ridley is the smallest sea turtle and a critically endangered species. Almost all females return each year to a single beach: Rancho Nuevo in Tamaulipas, Mexico.   © Michael Patrick O'Neill/OceanwideImages.com

Gregarious white pelicans prefer to gather in shallow coastal lagoons and lakes to feed.   *Photo:* Patricio Robles Gil

Coral reefs occur along the mid- to outer-edge of the continental shelf off the Big Bend area of Florida, and especially at the shelf break off of Texas and Louisiana—an area that includes the well-known Flower Gardens. Extensive areas of scattered banks and coral heads occur in very shallow shelf regions around south and central Florida. Shell reefs of the eastern oyster abound in the region.

Human Activities and Impacts

The northern Gulf is highly impacted by human activity—both directly and indirectly—and conservation issues are a major concern. A quarter of the commercial shipping of the US passes through the Straits of Florida, often causing damaging anchor and grounding scars in coral communities, and potentially introducing exotic species from ballast water. The human activity associated with the major petroleum hydrocarbon formations offshore in the northern Gulf coast is also a cause for concern: a vigorous complex of offshore petroleum exploration, extraction, shipping, service, construction, and refining industries has developed over the past half century, particularly in Louisiana and Texas, resulting in severe impacts on coastal wetlands, brine discharges, heavy metal deposition in drilling muds and tailings, and large- and small-scale petroleum discharges and major spills. Recently, coastal Florida has been the focus of potentially large expansion of these oil and gas-related activities. Within the past few years there has been significant interest in the development of liquefied natural gas (LNG) import facilities along US coastlines, particularly in the Gulf of Mexico. The proposed use of "open-loop," or once-through, systems for LNG regasification have raised concerns regarding the volume of water intake, generation of thermal plumes, discharge of treated water, increased turbidity, and generation of noise in the marine environment. In US Gulf of Mexico waters one open-loop terminal has recently been constructed, two others have been approved, and four more have pending license applications. Urban growth, shoreline development, freshwater reductions, and tourism also have affected the natural communities. Eutrophication in areas of high river discharges (and the expanding "dead zone" of oxygen-deprived and lifeless water caused in part by river-borne pollutants), reductions in fresh water inflow to estuaries due to upstream development and wetland loss due to subsidence and impoundment also place the region under duress. Harmful algal blooms and fishkills occasionally affect the area—in certain areas within the region the eastern oyster populations and seagrass communities have been affected by algal blooms and turbidity events. Overfishing and bycatch and habitat impacts from shrimp trawl fisheries are a concern. Two out of 17 major US federally-managed fish stocks are overfished, with 10 of unknown or unde-fined status (NMFS 2007). In areas influenced by freshwater, such as in Louisiana and Texas, crayfish harvesting forms an important economic base. Other conservation issues of major concern include protection of wading bird, shorebird, seabird, and endangered sea turtle populations.

14. Southern Gulf of Mexico

Level II seafloor geomorphological regions include:

Level III coastal regions include:

Regional Overview

Habitats of the southern Gulf of Mexico (hereafter referred to as the southern Gulf), such as coastal lagoons, estuaries, and dunes to mangroves, seagrass beds and some coral reefs help to support the 1,000 plus species of fishes that occur in the Gulf of Mexico. The region also supports oil and gas production, fisheries, and tourism. It spans the southern, tropical portion of the Gulf of Mexico (hereafter referred to as the greater Gulf), a semi-enclosed sea basin with tropical currents. Waters off the states of Veracruz, Tabasco, Campeche and Yucatan, as well as the lower portion of the slope off Florida and the Mississippi Fan, are included in this region. The boundary between the northern Gulf and the southern Gulf stems from seasonal changes in the distribution of some fish species due to surface water changes in the northernmost portion over the winter.

Physical and Oceanographic Setting

A prominent feature in the Gulf of Mexico is the Loop Current. This current, the major driving force bringing oceanic water into the greater Gulf, enters through the Yucatan Channel and exits through the Straits of Florida to become the Florida Current and later the Gulf Stream. Large, unstable rings of water are shed from the Loop Current, transporting massive amounts of heat, salt and water across the greater Gulf. Thus, the Loop Current plays an important role in shelf nutrient balance, at least in the eastern Gulf of Mexico.

The waters over the broad, shallow shelves in the eastern parts of the region are strongly wind-driven out to depths of approximately 50–60 m. The region is also extremely topographically diverse, and includes smooth slopes, escarpments, knolls, basins and submarine canyons. The southern coast of the greater Gulf has a considerably wide shelf (composed of shallow carbonate platform) towards its more eastern portion—up to 170 km in front of Campeche and up to 220 km at the northern coast of Yucatan. It narrows westward along the coast to between 6 and 16 km at its narrowest point in front of San Andrés Tuxtla. The adjacent slope in this region is steepest in the east (off the Yucatan peninsula), and gentlest in the west. A large part of the Mexico Basin is also found within the southern Gulf. The Sigsbee Abyssal Plain, with a depth of 3,600 m, is the deepest portion of the region. The southern Gulf's bottom type is a mixture primarily of sands (calcium carbonate), silt and clay.

Fact Sheet

Rationale: The semi-enclosed Gulf of Mexico is divided into northern and southern region by a sea surface temperature differential in winter.

Surface: 833,568 km²

Sea surface temperature: 24–25°C (winter), 28–28.5°C (summer)

Major currents and gyres: Loop Current

Other oceanographic features: continental shelf wind driven upwellings; cold fronts known as "nortes" during autumn, winter and spring; mixed and diurnal tidal regime; tropical currents

Physiography: semi-enclosed basin

Depth: shelf (roughly 0–200 m): 24%; slope (roughly 200–2,500/3,000 m): 33%; abyssal plain (roughly 3,000+ m): 43%

Substrate type: mixed sands, silt and clay

Major community types and subtypes: deltaic systems, coastal lagoons, estuaries, river inlets, nearshore and offshore coral reefs, mangroves, seagrass beds

Productivity: moderately high (150–300 g C/m²/yr)

Endemics: Maya octopus

Species at risk: manatee; Kemp's ridley, loggerhead, green, hawksbill, leatherback sea turtles; and the endemic Maya octopus. Species such as the Atlantic sharpnose shark, silky shark, blacktip shark and bull shark are presenting signs of over exploitation from artisanal shark fisheries targeted at immature individuals. Other species of concern include the whale shark, basking shark, great white shark, the large- and smalltooth sawfishes, and Atlantic and giant manta rays.

Important introduced and invasive species: deliberately introduced whiteleg shrimp

Key habitat: deltaic systems, coastal lagoons, estuaries, river inlets, near-shore and offshore coral reefs, mangroves, seagrass beds, volcanoes, brine pools and oil/gas seep biotic communities

Human activities and impacts: overfishing of all commercial species, port development, oil extraction and transportation, coastal pollution, coastal habitat destruction

Biological Setting

Productivity in the southern Gulf of Mexico ranges from eutrophic conditions in coastal waters to oligotrophic in the deeper ocean. Habitat and community type of the region also varies, and includes coastal lagoons, estuaries, dunes, mangroves, seagrass beds, as well as some coral reefs. Meteorological fronts occur seasonally and they are likely to have a significant impact on rates of primary production.

Upwelling along the western part of the Gulf of Campeche may lead to increased vertical inputs of nutrients, which in turn increase primary production and affect a wide array of associated species. There are most likely over 1,000 species of finfish in the Gulf of Mexico; however, only a small fraction of those have direct economic value and therefore are subject to exploitation. Reef fish in the region include groupers, snappers, amberjacks, scad and triggerfish. Shrimp, including brown, white, and pink, are also important for the region. The southern Gulf is also home to a number of species at risk, including manatee, sea turtles, such as Kemp's ridley, loggerhead, green, hawksbill and leatherback, as well as numerous shark species.

A spectacular diver, the masked booby is the largest of the boobies and spends the majority of its time at sea.
Photo: Patricio Robles Gil

Due to the great amount of freshwater runoff and river sediments, the lowlying coastline of the southern Gulf is largely free of coral formations. Coastal reef structures are present, however, near the cities of Tuxpan and Veracruz and offshore reef structures occur along the eastern portion of the Yucatan Shelf.

Human Activities and Impacts

The region's coasts are inhabited by large population centers, and natural resource use accounts for a major portion of the greater Gulf coast's economy. Overall, many upland, waterfront, and offshore activities—including tourism, commercial and recreational fishing, artificial reefs production, seafood production, boating, recreational activities on beaches and at marinas, shipping, petroleum production, and urban development—greatly impact the region. Shoreline alteration, pollutant discharge, oil and gas development, disease prevalence, alien invasive species, and nutrient loading are particular stresses of the region. The greater Gulf's bays, estuaries and coasts, in particular, are showing signs of stress that can be directly related to toxic chemicals, physical restructuring of the coast, local harvesting of preferred species such as shrimp, and nutrient loading (Birkett and Rapport 1999).

Infrastructure from oil and gas production—such as refineries, petrochemical and gas processing plants, supply and service bases for offshore oil and gas production units, platform construction yards, pipeline yards, and other industry-related installations—are found throughout the greater Gulf, and concentrated in southwestern Veracruz, Tabasco, and Campeche.

Commercial fishing is an important component of the Gulf's economic value. Traditional fisheries of the greater Gulf include penaeid shrimp and menhaden. In recent years, new fisheries such as those targeting reef fish, coastal migratory pelagic fish, and large oceanic pelagics have also sprung up. These fisheries have reached their harvesting limits. The brown, white and pink shrimp, Maya octopus (which is endemic to the continental shelf waters of the Yucatan Peninsula), red grouper and the brackish water clam, Atlantic rangia, are species of particular social and economic importance to Mexico. The red grouper fishery of the Campeche Bank, for example, is the second-most important exploited fish resource in Mexican waters of the greater Gulf and is particularly developed on the northern continental shelf of Yucatan. Red grouper is an important regional fishery and one of the few mono-specific grouper fisheries with trinational management (US-Mexico-Cuba). As a result of greater acceptance of shark meat in seafood markets, as well as the high price of shark fins in the oriental market, shark is also of high economic importance—catches of these species increased dramatically in the 1980s. Red snapper is apparently the most over-fished species in the Gulf of Mexico (Goodyear 1996). Scombroid fishes—including mackerel and tuna—are highly migratory species also found in the region. The status of their various stocks range greatly from healthy (Spanish mackerel) to fully utilized (yellowfin tuna) and overfished (king mackerel) to severely overfished (bluefin tuna) (NOAA 2002).

Flamingos are very social wading birds that inhabit mangroves, salt lagoons and coastal dunes.   *Photo:* Maciek Burgielski

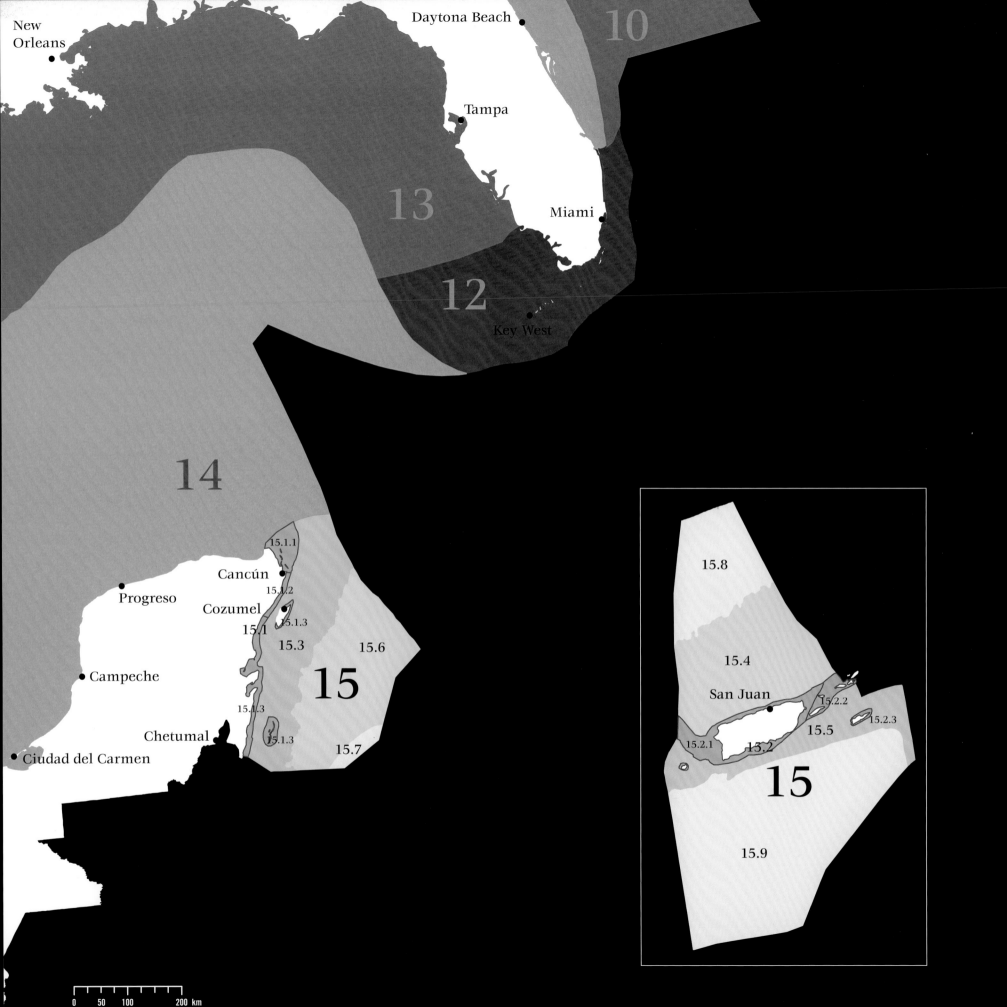

New
Orleans

Daytona Beach

10

Tampa

13

Miami

12

Key West

14

15.1.1

Cancún

Progreso

15.1.2

Cozumel

15.1.3

15.1

15.3

Campeche

15.6

15

15.1.3

Chetumal

15.1.3

15.7

Ciudad del Carmen

15.8

15.4

San Juan

15.2.2

15.2.1

15.5

15.2.3

15.2

15

15.9

0   50   100   200 km

15. Caribbean Sea

Level II seafloor geomorphological regions include:

15.1   Mesoamerican Caribbean Shelf
15.2   Insular Caribbean Shelf
15.3   Mesoamerican Caribbean Slope
15.4   Puerto Rico Trench
15.5   Insular Caribbean Slope
15.6   Yucatan Basin
15.7   Caribbean Caiman Mountain Range
15.8   Nares Abyssal Plain
15.9   Caribbean Basin

Level III coastal regions include:

15.1.1   Contoyan Neritic
15.1.2   Cancunean Neritic
15.1.3   Sian Ka'anean Neritic
15.2.1   Puerto Rican Neritic
15.2.2   US Virgin Islands/Viequean Neritic
15.2.3   St. Croix Neritic

Regional Overview

The Caribbean Sea Region region is a semi-enclosed tropical sea formed by the arc of the Greater and Lesser Antilles and the Atlantic coasts of Venezuela, Colombia, Central America and Mexico's Yucatan Peninsula. Waters off the State of Quintana Roo (Mexico), and US waters around the Commonwealth of Puerto Rico, the Territory of the US Virgin Islands, and Navassa Island[12] are included in this region. The Caribbean Sea Region is home to the Mexican extent of the Mesoamerican Caribbean Reef System. Coral reefs, along with the mangrove forests and seagrass meadows, provide important feeding and breeding areas for the more than 1,300 fish species, numerous marine mammals and the six sea turtles found in the region—many of which are endangered. The region's coral reefs are undergoing much stress—with some species suffering declines of over 90 percent throughout the Caribbean. Moreover, high value grouper and conch fisheries have collapsed in much of the Caribbean, and concern is increasing over the long-term sustainability of spiny lobster. Tourism—which is of great economic importance for the Caribbean—has developed intensely and rapidly, with often damaging ecological consequences.

Physical and Oceanographic Setting

The Caribbean Sea Region is a tropical, nutrient-poor sea that lies over primarily mixed sediments with increasing terrigenous components, especially in the western Caribbean. The principal surface water masses entering this region are from the North Brazil Current, around Trinidad, turning west along the continental slope into the southern Caribbean and the North Equatorial Current, passing into the Caribbean through the passages in the Lesser and Greater Antilles. The major flow of the Caribbean Current passes around the southern part of the Caribbean toward the Yucatan Channel, through which water leaves the Caribbean and enters the Gulf of Mexico.

Due to the size of the Caribbean Sea and its wide latitudinal range, variations in climate exist. The region is characterized by strongly seasonal rainfall patterns and stochastic, large-scale disturbances, in the form of tropical storms and hurricanes. The eastern Caribbean tends to be dry-tropical climate, in contrast with the western

---

12   Navassa Island is a small, uninhabited oceanic island and US protectorate located between Jamaica and Haiti, 50 km off the southwest tip of Haiti. It is under the jurisdiction of the US Fish and Wildlife Service as one component of the Caribbean Islands National Wildlife Refuge. Based on some preliminary quantitative surveys in 2000, and because of its isolated and uninhabited status, Navassa has been presumed to include relatively pristine reef ecosystems. Local land-based anthropogenic pollution and recreational uses are essentially absent. However, there is substantial but unquantified fishing activity at Navassa by transient Haitians and their impact has been suggested to be substantial and potentially rapidly increasing.

Fact Sheet

Rationale: a semi-enclosed sea distinguished by warm sea surface temperatures, clear oligotrophic waters, and numerous islands and banks

Surface: 306,138 km²

Sea surface temperature: 25.5°C (winter), 28°C (summer)

Major currents and gyres: Caribbean Current

Other oceanographic features: mixed tidal regimes, nutrient poor tropical sea

Depth: shelf (roughly 0–200 m): 6%; slope (roughly 200–2,500/3,000 m): 36%; abyssal plain (roughly 3,000+ m): 58%

Substrate type: mixed sands (calcium carbonate)

Major community type and subtype: coral reefs, mangroves, seagrass beds, deepsea communities

Productivity: moderately high (150–300 g C/m²/yr) in nearshore habitats such as coral reefs, mangroves and seagrass beds; oligotrophic conditions elsewhere, reflecting considerable spatial and seasonal heterogeneity in productivity throughout the region

Endemics: goby (Gobiidae)[13]

Species at risk: Roseate tern; loggerhead, green, hawksbill, and leatherback sea turtles; manatee; black coral; elkhorn and staghorn coral; queen conch; Nassau grouper. Species such as the Atlantic sharpnose shark, silky shark, blacktip shark and bull shark are presenting signs of over exploitation from artisanal shark fisheries targeted at immature individuals.

Important introduced and invasive species: deliberately introduced whiteleg shrimp, big pink jellyfish

Key habitats: coral reefs, seagrass beds, mangrove forests

Human activities and impacts: habitat loss from intensive coastal tourism, urbanization, land-based sources of pollution, artisanal fisheries

13 See http://www.coralreeffish.com/gobiidae.html. There are over a hundred species of gobies in Caribbean waters.

Caribbean that tends to be wet-tropical with large sediment input from rivers. The Caribbean Sea is adjacent to, and influenced by two enormous river systems to the southeast (the Orinoco River at the eastern margin, and the Amazon River south of that).

The Mexican section of the Caribbean along the Yucatan Peninsula is the northern extent of the Mesoamerican Caribbean Reef System—the second-largest reef system in the world—which stretches from the northern tip of the Yucatan Peninsula to the Bay Islands off the coast of Honduras. The Mexican portion of the region has a narrow continental shelf with a width of 20 km around Cancun, and 1 to 3 km in the Sian Ka'an region, and also includes the islands of Cozumel, Mujeres and Contoy as well as Banco Chinchorro off the continental shores. Its waters are also influenced by underground freshwater sources that have created vast underground channels and caves and cenotes on the surface in the Yucatan and Sian Ka'an. regions The continental margin off the Mexican portion is extremely complex, beginning with a gentle slope that turns to a very steep escarpment before reaching the Yucatan Basin. The southern portion of the deep Yucatan Basin (over 4,500 m) borders the Caribbean Caiman Mountain Ridge—a submarine mountain chain that rises over 4,000 m from the sea bottom, producing banks at depths of less than 200 m below sea level. The US portion of the Caribbean

(Puerto Rico and the Virgin Islands Bank, and St. Croix) represents a more insular part of this Caribbean system. The island of Puerto Rico is the fourth-largest in the Caribbean. It shares a shallow shelf with several small islands (e.g., Culebra and Vieques—Puerto Rico, USA) as well as St. John and St. Thomas (US Virgin Islands), and the British Virgin Islands. A deep channel separates the main island of Puerto Rico from the small island of Mona to the west; another deep channel to the east isolates Puerto Rico and the US Virgin Islands from St. Croix. About 120 km to the north, the Puerto Rico Trench—the deepest submarine depression in the North Atlantic Ocean—borders Puerto Rico. The Puerto Rico Trench is about 1,750 km long, 100 km wide, and reaches a depth of 8,380 m.

Biological Setting

Although upwelling along the northern coast of Colombia contributes to relatively high productivity in small areas of the southern Caribbean, the remaining waters of the Caribbean are mostly clear and nutrient-poor, except in coastal areas with high river sediment and nutrient input (e.g., Gulf of Honduras). Examples of this in the Mexican Caribbean can be found in Ascensión and Espiritu Santo Bays within the Sian Ka'an Biosphere Reserve, where freshwater and nutrient inputs result from sloughs and sheet flow over coastal wetlands. Key components of the shallow water ecosystem are the seagrass beds, sand banks and muddy bottoms. Coral reefs in the Caribbean Sea are important for their biological diversity and unique ecological processes, but make up a relatively small area of the total shallow-water benthic habitat. Living corals are mostly calcium-secreting corals, that thrive in clear, oceanic, shallow, low-nutrient waters, with plenty of sunlight and warm temperatures. Sponges are very species-rich and common in the Caribbean on coral reefs and seagrass beds, where they play an important role in maintaining water clarity and in secondary production.

Coral reefs, mangrove forests and seagrass meadows form large coastal systems or complexes that can provide important habitat—such as feeding and breeding areas for the more than 1,300 fish species, numerous marine mammals and sea turtles found in the region. Mangroves also provide additional environmental services, such as erosion control, nutrient retention, and storm buffering.

The region is home to many species at risk, such the loggerhead, green, hawksbill and leatherback sea turtles; manatee; black coral; giant manta ray; Nassau grouper and many shark species, as well as to the overexploited queen conch. The region has recently experienced massive die-offs of the common reef-building corals, elkhorn coral and staghorn coral.

Fringing reefs are common adjacent to small islands and cays in the Caribbean. In the Mexican portion of the Mesoamerican Reef System, a barrier reef system and fringing reefs are also common. Offshore atolls and banks are also present. These reefs represent a global biodiversity conservation priority.

The islands of the Antilles, including the US Caribbean, are typically endowed with leeward reefs on their coasts as well as fringing areas of mangroves and small seagrass beds. Reefs in the Western Atlantic are taxonomically much less diverse and lack many of the characteristic biota of Indo-Pacific reefs, but still represent the most diverse Atlantic marine ecosystems. Unfortunately, the deeper habitats in the Caribbean are poorly explored, but endemic species have been discovered (e.g., fishes of the *Gobiidae* family).

Human Activities and Impacts

The Caribbean Sea is showing signs of stress, particularly in the shallow waters of coral reefs. Coral growth can be limited by a number of factors, including high turbidity, exposure to fresh water or air, extreme temperatures, pollution, and excess nutrients. Major human stresses on coral reefs of the region include climate change, coastal development and runoff, coastal pollution, as well as overfishing and other detrimental fishing practices.

Caribbean spiny lobsters are noturnal creatures and hide under ledges and in crevices during the day.
Photo: Claudio Contreras

A French angelfish sheltering next to an elkhorn coral.
Photo: Claudio Contreras

The region is also subject to coral bleaching and mortality events, which increased during the 1980s and 1990s. Bleaching—which occurs when the coral expels its symbiotic algae and, if prolonged, may result in coral mortality—has generally been linked to abnormally high water temperatures in the Caribbean in recent years. In 1983, an unknown disease swept through the Caribbean, causing mass mortality of the long-spined sea urchin. In conjunction with overfishing, loss of this keystone herbivore has led to overgrowth of corals by macroalgae on many reefs. Shallow water elkhorn and staghorn coral have also suffered declines of over 90 percent throughout the Caribbean, primarily as a result of white band disease.

Moreover, overfishing has occurred in this region on nearly all reefs near inhabited islands or coasts. Over 170 species are caught for commercial purposes in the Caribbean Sea, but most of the catch is actually of fewer than 50 species. The principal species harvested are coral reef fishes, conch and Caribbean spiny lobster. High value grouper and conch fisheries have collapsed in many Caribbean areas. Due to an increase in fishing effort, there is also concern over the long-term sustainability of spiny lobster—one of the most valuable species of the Caribbean. Coral reef fisheries are mostly small-scale, artisanal fisheries, but are of significant economic, cultural and recreational importance.

The nearshore marine resources, especially coral reefs, mangroves and seagrass beds, have been degraded by natural and anthropogenic impacts. Because of its pleasant and agreeable climate, the Caribbean has been subject to intense and rapidly growing tourism year-round—attested to by the intensive coastal development and number of cruise ships in the area. Unfortunately, with this tourism have also come poorly planned coastal land development, point and non-point source pollution, conversion of mangrove habitats and upland deforestation, unsustainable fisheries, and recreational overuse—all of which have contributed to the area's environmental degradation. Over the last 25 years, for example, the state of Quintana Roo has experienced exponential growth derived from tourism activities, including coastal development and cruise ships. Widespread inland deforestation and pollution of the water table within a karstic system (underground rivers through limestone caves and channels) are also of major concern for Quintana Roo.

Tourism, much of it marine oriented, is of major economic importance to both Puerto Rico and the US Virgin Islands—in 2000, Puerto Rico and the US Virgin Islands hosted 4.57 million and 2.5 million visitors, respectively, many of them cruise ship passengers. This is over and above the already concentrated population found in Puerto Rico of 3.8 million people (2000)—one of the most densely populated islands in the world—and in the US Virgin Islands of roughly 110,000 people. Puerto Rico and the US Virgin Islands have witnessed increased incidences of hurricane damage and coral bleaching in the last two decades. Unless carefully managed, increased tourism expected for the region will contribute to more environmental degradation—an awful fate for this fragile and critically important region. The use of MPAs as a management tool for improving the condition of the Caribbean's nearshore ecosystems is a reason for hope for this otherwise grim situation.

The cormorant is one of the numerous species of seabirds that one encounters in Sian Ka'an Biosphere Reserve, in the Mexican Caribbean. *Photo:* Hans Hillewaert

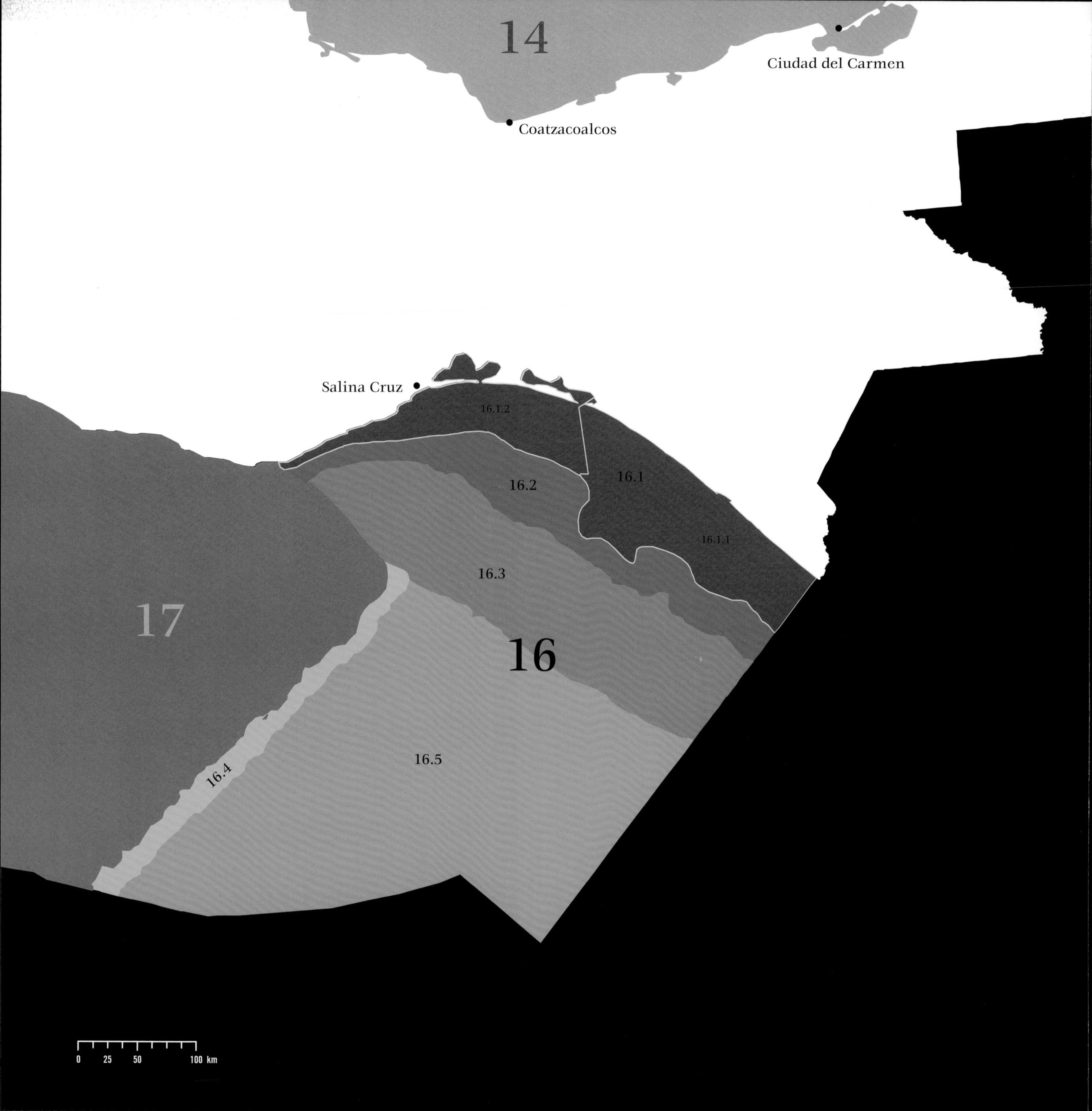
14
Ciudad del Carmen
Coatzacoalcos
Salina Cruz
16.1.2
16.1
16.2
16.1.1
16.3
17
16
16.4
16.5
0 25 50 100 km

16. Middle American Pacific

Level II seafloor geomorphological regions include:

16.1	Tehuantepec Gulf Shelf
16.2	Tehuantepec Gulf Slope
16.3	Mesoamerican Trench
16.4	Tehuantepec Ridge
16.5	Guatemala Basin

Level III coastal regions include:

| 16.1.1 | Tehuantepecan/Chiapan Neritic |
| 16.1.2 | Tehuantepecan/Oaxacan Neritic |

Regional Overview

The Middle American Pacific Ecoregion—largely free from mixing with colder waters from farther north and therefore described as a year-round tropical sea—supports important fisheries such as yellowfin and skipjack tuna, as well as shrimp. Bycatch, however, is of great concern—the region's bycatch ratio is the highest in Mexico. Although relatively small, the region's bathymetry is quite diverse and includes a relatively wide continental shelf that drops off to the continental slope and Mesoamerican Trench, and rises to the Guatemala Basin and the Tehuantepec Ridge. Waters off the Mexican states of Oaxaca and Chiapas are included in this region.

Physical and Oceanographic Setting

The Middle American Pacific remains essentially free from the southernmost winter influence of the California Current year round and thus is a tropical sea. The Mexican portion of the Middle American Pacific, encompassing the Gulf of Tehuantepec and adjacent waters, is influenced by the Costa Rica Current (derived from the North Equatorial Counter Current). It is also a region that experiences high seasonal variability due to upwelling, and is strongly influenced by freshwater discharge from coastal

lagoons and river systems present in coastal areas in Chiapas, as well as winds from the Gulf of Mexico. Because the Gulf of Tehuantepec lies south of a major gap in the Central American sierra chain, transmontane winds (called "Tehuanos") from the Gulf of Mexico[14] have easy passage. The northern winds force the surface airflow from the Gulf of Mexico towards the Gulf of Tehuantepec. Interaction between this and the northward flowing Costa Rica Coastal Current generates a meridional thermocline (pycnocline and nutricline) ridge. During the winter months, the layer above this feature may be completely mixed by wind stress, and surface temperature, salinity, and nutrient levels resemble values in the pycnocline. After extreme transmontane wind events, a plume having such characteristics may stretch several hundred kilometers to the southwest from the Gulf. Surface productivity is high.

The Middle American Pacific includes a moderate to narrow continental shelf that widens towards the southeast; a continental slope with varied grades from gentle to steep; part of the Mesoamerican Trench—a subduction zone with steep slopes and profound depths (6,000 m); the eastern portion of the Guatemala

---

14   Only 200 km of plains barely broken by the hills and low rises of the Sierra Atravesada separates the Atlantic and Pacific Oceans at the Isthmus of Tehuantepec.

Fact Sheet[15]

Rationale: largely free of the southernmost winter influence of the California Current and therefore described as a year-round tropical sea

Surface: 148,380 km²

Sea surface temperature: 26–27°C (winter), 29.5°C (summer)

Major currents and gyres: North Equatorial Countercurrent/Costa Rica Coastal Current

Other oceanographic features: mesotidal, mixed, semidiurnal tidal regime; year round tropical sea. From October to May, the Tehuantepecanos winds (northern winds, more evident during winter) determine the upwelling, which cause high seasonal variability.

Depth: shelf (roughly 0–200 m): 16%; slope (roughly 200–2,500/3,000 m): 30%; abyssal plain (roughly 3,000+ m): 54%

Substrate type: silt, mud, sand

Major community types and subtypes: coastal lagoons, mangroves, sandy shore and benthic communities

Productivity: high (>300 g C/m²/yr), due to the equatorial upwelling, open ocean and coastal upwellings, and nutrient inputs coming from river run-off along the tropical areas

Endemics: the graceful herring, the Mexican clingfish; and the green alga *Codium oaxacensis*

Species at risk: loggerhead, East Pacific green, leatherback, and olive ridley sea turtles; purpura conch

Important introduced and invasive species: the brown alga *Sargassum muticum*

Key habitat: coastal lagoons, mangroves, and coral reefs

Human activities and impacts: pollution from agricultural runoff and urban activities, and overfishing of commercial species. A highly economically important artisanal shark fishery in the Gulf of Tehuantepec, and particularly in Chiapas, concentrates on the silky and hammerhead sharks.

15 The description refers only to the Mexican portion of this region.

Basin that is undulated with trenches that reach depths of 4,600 to 4,900 m; as well as the Tehuantepec Ridge—a submarine mountain chain of volcanic origin. The region's bottom type varies from mixtures of mud, sand and gravel, and the zone is characterized by shallow waters with minimal oxygen.

Biological Setting

Available data suggest that at least during the northerly windy season, the Gulf of Tehuantepec acts as a nutrient and phytoplankton-carbon pump, enriching adjacent offshore waters. When the Gulf behaves as a tropical ecosystem, there is low phytoplankton biomass and low primary productivity.

The most evident interannual variation is the effect of ENSO events, associated with a deep thermocline in the whole region, even during winter, as well as very low chlorophyll concentrations.

Many of the region's communities are characteristic of those found in upwellings. At least 153 species of marine algae have been found on the seafloor of the Gulf of Tehuantepec. At least 123 families with 172 genera and 239 species constitute the benthic invertebrate community of the Gulf. At least 178 species in 103 genera and 52 families constitute the demersal fish community. The highest diversity is found offshore of the estuarine systems during the rainy season. Mangrove communities are also found in the region and are more developed in Chiapas than in Oaxaca. The Oaxacan coast presents limited coral reef structures (in Bahia de Huatulco, La Entrega and Puerto Angel) in relatively good condition.

The loggerhead, East Pacific green, leatherback and olive ridley sea turtles, as well as other species at risk, such as the purpura conch and several shark species, use the waters of this region as their home. Endemic species are also found in the region, and include fishes such as the graceful herring and the Mexican clingfish, as well as the alga *Codium oaxacensis*.

Human Activities and Impacts

Fishing and coastal industrial development play important roles in the economy of the Middle American Pacific. This region supports important fisheries, such as that of yellowfin and skipjack tunas, as well as shrimp. Intense artisanal fisheries are found in the coastal lagoons, particularly in Chiapas and more recently in Oaxaca. Over the continental shelf, industrial shrimp trawlers sweep the sea bottom, killing juveniles of countless untargeted species (i.e., bycatch). Shrimp yields have been sharply declining over the last decades—a phenomenon that is creating conflicts between fishers in the region. The region's bycatch ratio is also the highest in Mexico (1:16 to 1:41, versus 1:12 in the Gulf of Mexico, and usually 1:10 in the Mexican Pacific) (Tapia-Garcia and Garcia-Abad 1998). Moreover, coastal industries and activities based on oil, sugar and transportation are also placing pressures on the Middle American Pacific. Several areas within the region are showing signs of critical pollution problems: the Salina Cruz port and surrounding areas contain oil and heavy metal pollution; in the Ventosa Bay and Estuary, oil and heavy metal pollution is compounded by the presence of fecal coliforms; and in Laguna Superior, pollution from pesticides, herbicides, oil and domestic discharges is prevalent (Tapia-Garcia *et al.* 1998).

The roseate spoonbill inhabits mangrove swamps and coastal lagoons. Along Mexico's Pacific coast it is found from Sonora down to the border with Central America. The species owes its name to the characteristic form of its large beak: flattened and rounded at the end, like a big spoon or spatula   *Photo:* Darlene F. Boucher

Juvenile spottail grunts seek refuge and food among the roots of a mangrove.
Photo: Octavio Aburto

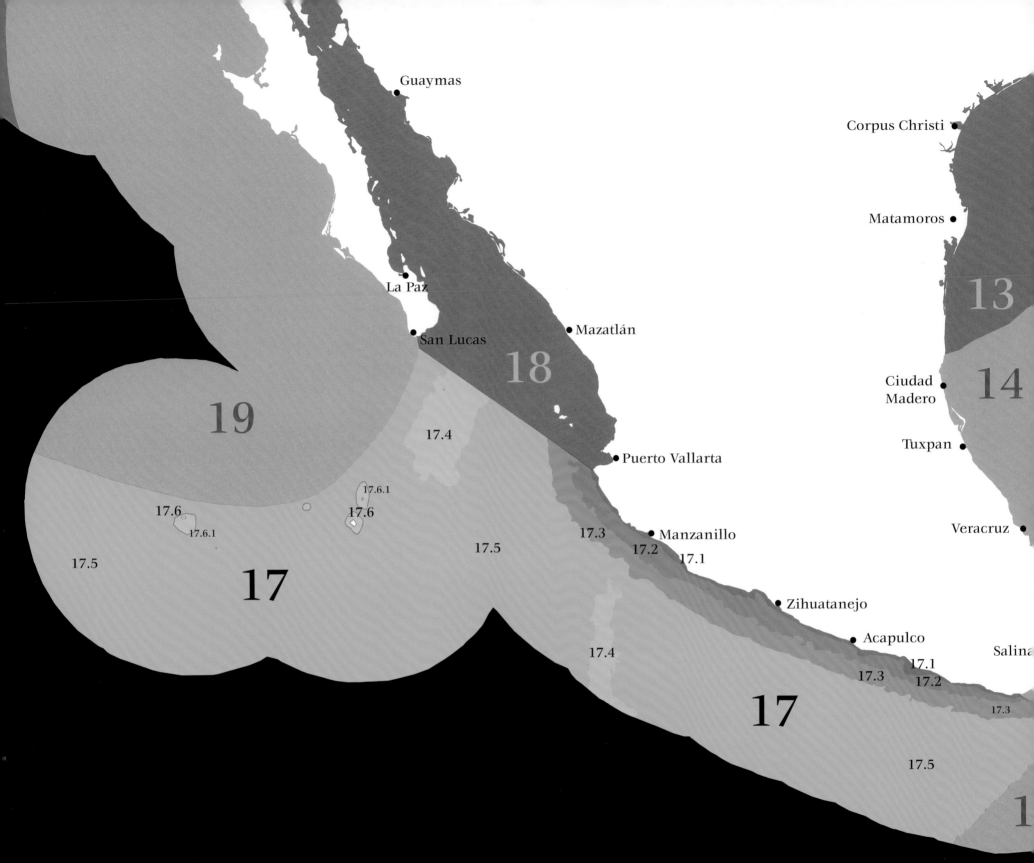

Guaymas

Corpus Christi

Matamoros

13

La Paz

San Lucas

Mazatlán

18

14

Ciudad
Madero

19

17.4

Tuxpan

Puerto Vallarta

17.6.1

17.6

Veracruz

17.6
17.6.1

17.3

Manzanillo

17.5

17.5

17.2

17.1

17

Zihuatanejo

Acapulco

Salina

17.4

17.1

17.3

17.2

17

17.3

17.5

1

0   50   100   200 km

17. Mexican Pacific Transition

Level II seafloor geomorphological regions include:

17.1   Mexican Pacific Transition Shelf
17.2   Mexican Pacific Transition Slope
17.3   Mesoamerican Trench
17.4   East Pacific Rise
17.5   Mexican Pacific Transition Plain and Seamounts
17.6   Revillagigedo Oceanic Archipelago

Level III coastal regions include:

17.1.1   Mexican Pacific Transition Neritic[16]
17.6.1   Revillagigedo Neritic

Regional Overview

The Mexican Pacific Transition is a fairly complex region, with a narrow shelf that drops off steeply to great ocean depths close to the coast. It is incised by several canyons and the Meso-american Trench that plunges to depths between 4,000 and 5,000 m. In addition, the region is dotted by numerous submarine hills and mountains, and includes a rift system and volcanic cones that have emerged from the depths of the ocean. It also has a great diversity of coastal systems and subsequently high species diversity. Tourism has contributed to shaping many of the coastal communities in the region. The Mexican Pacific Transition is a tropical sea that is seasonally affected by the winter influence of the California Current. Waters off the Mexican states of Jalisco, Colima, Michoacan, Guerrero and Oaxaca and the southernmost tip of Baja California Sur are included in this region. The northernmost limit of this region, in general coincides with where the California Current turns westward in summer, leaving the region under the influence of the warm Costa Rica Coastal Current.

Physical and Oceanographic Setting

The Mexican Pacific Transition encompasses the southernmost winter influence of the California Current, which in effect seasonally transforms this tropical sea into a subtropical one. The North Equatorial Countercurrent flows above the north slope of the equatorial thermocline ridge, from 120° E across the ocean to Central America where it turns northward and becomes the Costa Rica Coastal Current. Here it flows along the coast of Central America and mainland Mexico and meets with the California Current at the most northerly part of region 17. At this point it separates from the coast to feed the North Equatorial Current. The region is affected by hurricanes, which cause disturbances in the physical, biological and human systems of the region.

The Mexican Pacific Transition is a fairly complex region in terms of its geomorphology. The region's narrow continental shelf, which generally has a width of 10 to 15 km, drops off steeply so that great ocean depths (2,500 to 3,000 m) are reached extremely close to the coast. The shelf is incised by several canyons from Jalisco to Oaxaca. There are many coastal lagoons, mainly in the state of Guerrero, where the inlets of coastal lagoons are ephemeral—opened during the rainy season, and closed during the dry

---

16   This level III region—a very narrow fringe composed of rocky shores, sandy beaches, pocket beaches, coastal lagoons, deltas and estuaries—is not shown on the map due to reasons of scale.

Fact Sheet

Rationale: encompasses the southernmost winter influence of the California Current; its northern-most limit, in general coincides with where the California Current turns westward in summer, leaving the region under the influence of the warm Costa Rica Coastal Current

Surface: 1,038,010 km²

Sea surface temperature: 25–28°C (winter), 29.5°C (summer)

Major currents and gyres: North Equatorial Countercurrent/Costa Rica Coastal Current, California Current

Other oceanographic features: mesotidal, mixed, semidiurnal tidal regime; tropical sea with the southern-most winter influence of the California Current

Physiography: narrow continental shelf with amplitude of 10–15 km and a slope of less than 1° 30′, deep oceanic trench and complex abyssal plains

Depth: shelf (roughly 0–200 m): 2%; slope (roughly 200–2,500/3,000 m): 9%; abyssal plain (roughly 3,000+ m): 89%

Major community type and subtype: rocky shores, sandy shores, coastal lagoons, estuaries, generally small deltaic systems, reef patches and mangroves, deep-sea communities

Productivity: high (>300 g C/m²/yr)

Species at risk: loggerhead, East Pacific green, leatherback and olive ridley sea turtles. Species such as the silky shark, scalloped hammerhead, and smooth hammerhead are presenting signs of over exploitation from artisanal shark fisheries targeted at immature individuals. Other species of concern include the whale shark, great white shark, sawfish, and mantas (giant manta, spinetail mobula, smoothtail mobula, pygmy devil ray, and sicklefin devil ray).

Important introduced and invasive species: brown algae

Key habitat: estuaries, mangroves, coral communities, coral reefs. The main nesting beaches for the East Pacific green sea turtle are found on the beaches of Colola and Maruata in the state of Michoacán.

Human activities and impacts: fisheries, coastal tourism and urban development. Offshore fisheries may be impacting species with low fertility rates such as the pelagic thresher and bigeye thresher.

season. In general, the substrate on the shelf is characterized by a gradation of sand, muddy sand, sandy mud and mud from the coastline. Variations on this pattern may be related to anomalous high storms or to extremely high tectonic activity, as observed in the Guerrero state shelf.

The Mesoamerican Trench, continuous from the southern Middle American Pacific Ecoregion, has a depth of between 4,000 to 5,000 m in this region, and is located about 100 km offshore. The trench is a strong subduction zone that stretches along the coast-line and triggers frequent earthquakes in the region. Seaward of the trench, the region's abyssal plain is 3,500 to 4,000 m deep, and is marked by numerous submarine hills and mountains that rise 1,000 m above the seafloor, as well as by various fractures that can reach 4,900 m of depth. The region also contains a rift system where tectonic plates spread to create new seafloor, as well as the Revillagigedo Archipelago—tops of volcanic cones that have emerged from the depths of the ocean.

Biological Setting

The Mexican Pacific Transition is a region of high productivity. As a result of warmer sea temperatures, it supports a very different tropical marine fauna from that supported by the California and the Humboldt Currents. Many fish species are similar to the Panamanian fauna (to the region's south), with influence from

Mangroves and lagoons of Jalisco on the Pacific Coast.   *Photo:* Patricio Robles Gil

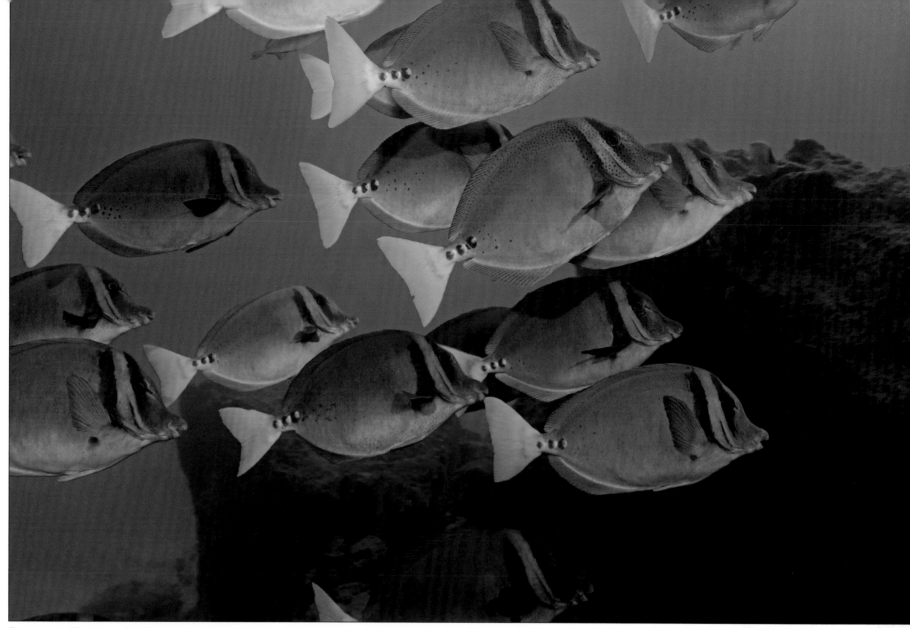

The razor surgeonfish, seen here off Isla Socorro in the Revillagigedo Archipelago, lives in large schools; juveniles are completely yellow. *Photo:* Octavio Aburto

the Gulf of California and the Southern Californian Pacific Ecoregions. As a result of the great diversity of coastal ecosystems (e.g., estuaries, coastal lagoons, coral communities, mangroves, rocky and sandy shores) there is high species diversity. For fishes alone, over 215 species exist in the region: speckled guitarfish, Chilean round ray, shorthead lizardfish, Pacific flagfin mojarra, burrito grunt, oval flounder, and the bullseye puffer are abundant and dominant. Of the fish species found on the continental shelf, 42 percent are estuarine dependent or associated to estuarine processes. A great abundance and dispersion of demersal fish communities can be found along the external shelf, towards 100 m depth. El Niño (ENSO) event affects coastal benthic/demersal fishes in the region.

Diverse and abundant coral communities exist near Zihuatanejo (Guerrero) and the Revillagigedo Archipelago, whose coral and rocky reefs house many endemic species. The dominant species of coral are *Pocillopora damicornis* and *P. verrucosa*. These coral communities also support high fish diversity, with more than 120 species. Species such as Panamic sergeant major, threebanded butterflyfish, Acapulco damselfish, giant damselfish, razor surgeonfish, and sunset wrasse are abundant and dominant. There are also more than 189 species of mollusks. Gastropod mollusks of commercial interest (in the genera *Fusinus, Hexaplex, Ficus, Harpa, Bursa and Cantharus*) from the shelf of Jalisco and Colima are associated with sandy silt and medium sand substrate in the region.

The region also presents very important nesting beaches for loggerhead, East Pacific green, leatherback and olive ridley sea turtles. The Mexican states of Michoacán (Mexiquillo, in particular), Guerrero and Oaxaca once supported what was at one time the largest nesting population of leatherback sea turtles in the world (following the decline of the Malaysian population). The population, however, has declined precipitously in the last decade. The region still houses the main nesting beaches for the East Pacific green sea turtle—at the beaches of Colola and Maruata. Although these beaches are said to hold approximately one-third of the population, they too are being threatened, and the population has exhibited a clear decline over the past 40 years (NMFS and USFWS 1998).

Human Activities and Impacts

Increased population pressures on the Mexican Pacific Transition coast have led not only to overfishing, but also to pollution of rivers, streams, lakes and coastal lagoons. The most valuable fisheries of the region are offshore tuna—yellowfin, bigeye and skipjack—captured by industrial fleets. Artisanal finfish and shark fisheries operate all along the coast, and shrimp is also an important fishery for the region. In addition to commercial fishing, sport-fishing for billfish (e.g., Indo-Pacific blue and stripped marlins, swordfish, sailfish) represents an important economic activity to the region as well. The purpura conch, which is collected and used in fabric dying, is a species at risk locally. Finfish populations have been experiencing declines due in part to shark fisheries bycatch; populations of sharks have also diminished. Habitats vital for fisheries, such as estuaries and mangroves, are increasingly subject to pollution and are being rapidly destroyed. The lack of facilities for proper solid waste disposal and the lack of basic infrastructure create a variety of challenging situations for the region. Moreover, runoff of fertilizers from agricultural lands also contributes to the degradation of the region's coastal waters by increasing the risk of toxic algal blooms. These discharges are thought to compromise the shellfish harvests as well as endanger the health of bathers using the waters around the main ports (NOAA 2002). Coastal tourism is another factor that has contributed to shaping the form of the Mexican Pacific Transition—communities such as Puerto Vallarta (Jalisco), Acapulco and Zihuatanejo (Guerrero), Huatulco (Oaxaca), and Manzanillo (Colima), as well as other coastal communities to a lesser extent, have had massive tourism development. An industrial port with steel mills has been developed in Lázaro Cárdenas, Michoacán.

The olive ridley sea turtle laying eggs at Escobilla, Oaxaca.
Photo: Patricio Robles Gil

The rugged, rocky coastline of Guerrero. *Photo:* Patricio Robles Gil

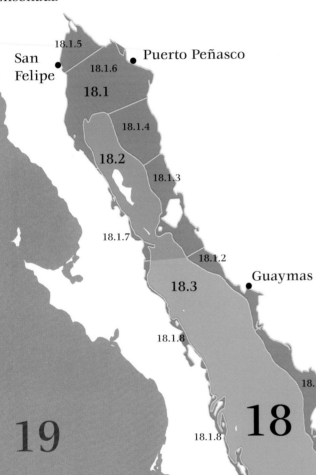

Los Angeles

Ensenada

18.1.5

San
Felipe

Puerto Peñasco

18.1.6

18.1

18.1.4

18.2

18.1.3

18.1.7

18.1.2

Guaymas

18.3

18.1.8

18.1.1

19

18

18.1.8

18.1.9

La Paz

18.1

18.1.9

18.3

Mazatlán

San Lucas

18.5

18.4

18.1.1

17

18.6

Puerto Vallarta

0   50   100   200 km

18. Gulf of California

Level II seafloor geomorphological regions include:

18.1   Cortezian Shelf
18.2   Midriff Islands Straits
18.3   Gulf of California Slope and Basins
18.4   Gulf of California Plains and Seamounts
18.5   East Pacific Rise
18.6   Mesoamerican Trench

Level III coastal regions include:

18.1.1   Eastern Cortezian Neritic
18.1.2   Guaymean Neritic
18.1.3   Tiburonian Neritic
18.1.4   Loboian Neritic
18.1.5   Upper Cortezian Inner Neritic
18.1.6   Upper Cortezian Outer Neritic
18.1.7   Northern Baja Californian Neritic
18.1.8   Southern Baja Californian Neritic
18.1.9   Cape/Cortezian Neritic

Regional Overview

The Gulf of California (also known as the Sea of Cortez or Mar de Cortés) is a semi-enclosed sea known for its exceptionally high levels of biodiversity and rates of primary productivity due to a combination of its topography, warm climate, and upwelling systems. It is also home to the endemic vaquita porpoise—the most endangered cetacean in the world—and the large, corvina-like totoaba. Upstream damming and diversion leading to decrease of fresh water input from the Colorado River has drastically changed the ecological conditions of the Upper Gulf—now a hypersaline estuarine system important for fish reproduction. Fishing, especially with gillnets, is a key activity for coastal communities of the region. However, decreases in abundance of several species of fish and changes in gear types have caused much concern. Waters off the Mexican states of Nayarit, Sinaloa, Sonora, Baja California and Baja California Sur are included in this region. The southern border of this region is generally considered, oceanographically and faunally, to stretch from Cabo Corrientes (at the northwestern tip of the state of Jalisco) on the mainland to Cabo San Lucas at the tip of the Baja California peninsula (Brusca *et al.* 2005).

The region includes five B2B Marine Priority Conservation Areas (PCAs): PCA 24-Corredor Los Cabos/Loreto; PCA 25-Alto Golfo de California; PCA 26-Grandes Islas del Golfo de Califonia/Bahía de Los Ángeles; PCA 27-Humedales de Sonora, Sinaloa y Nayarit/Bahía de Banderas; and PCA 28-Islas Marías (Morgan *et al.* 2005).[17]

Physical and Oceanographic Setting

The Gulf of California is a long, narrow, semi-enclosed sea (approximately 1,000 km long and 150 km wide, spanning over nine degrees of latitude). It is bordered by the Mexican coastal states of Sonora, Sinaloa, and Nayarit to the east, the two states of the Baja California peninsula to the west, and the Colorado River delta (Sonora/Baja California) to the north. It is an exclusively Mexican sea, once influenced by a US watershed (the Colorado River). Deep basins (greater than 3,000 m at the entrance to the Gulf), steep slopes, both narrow and wide continental shelves, numerous islands, sandy bays and beaches, and coastal lagoons (mostly hypersaline) characterize the region. The Guaymas Basin has tectonic activity

---

17   See footnote 8 (p. 11).

Fact Sheet

Rationale: a semi-enclosed sea.

Surface: 265,894 km²

Sea surface temperature[18]: 13–21°C (winter), 28–31°C (summer)

Major currents and gyres: upwelling on the east coast during winter and spring and on the west coast during the summer. Large and complex temporal eddies can occur year round, however

Other oceanographic features: semi-enclosed subtropical sea with high seasonal variability and exceptionally high primary productivity. Three layer circulation: the direction of transport of the surface layer changes seasonally with the large-scale winds; strong tidal currents and convective overturn during winter in the northern Gulf (between the mouth of the Colorado River and Tiburon and Angel de la Guarda Island); the Gulf's entrance (a triangular area between Cabo San Lucas, Mazatlán and Cabo Corrientes) has a very complicated thermohaline structure characterized by fronts, eddies and intrusions which can be linked to the confluence of three distinct currents; tidal regime is mixed and semidiurnal at the mouth of the Gulf, diurnal in its mid section and semidiurnal in the northern Gulf; tidal amplitudes vary from 1 to 7 m.

Physiography: spreading rift system; very wide continental shelf in the northern Gulf, medium-width shelf with abundant coastal lagoons in its eastern portion and island-studded narrow shelf in the western Gulf

Depth: shelf (roughly 0–200 m): 32%[19]; slope (roughly 200–2,500/3,000 m): 54%; abyssal plain (roughly 3,000+ m): 14%

Substrate type: mixed sands, silt, mud, rocky reefs and rhodolith beds

Major community type and subtype: coastal lagoons, deltaic systems, mangroves, seagrass beds, rocky shores, sandy beaches, coral reefs, hydrothermal vents and rhodolith beds

Productivity: High (>300 g C/m²/yr), one of the most productive marine ecosystems in the world. The northern Gulf has two main natural fertilization mechanisms: tidal mixing around the large islands and the wind-induced upwelling along the east-central gulf.

Endemics: About 10 percent of the Gulf of California fish fauna is endemic—80 of the 821 bony fish and 4 out of the 90 cartilaginous fish species are endemic. These include totoaba, Gulf weakfish, bigeye croaker, Cortez clingfish, whitetip reef shark, Cortez stingray, Cortez skate, Gulf grunion, delta silverside, leopard grouper, sawtail grouper, Cortez damselfish, Gulf signal blenny, Sonora blenny, slow goby, Guaymas goby, and shortjaw mudsucker, among others. Mammals include the vaquita porpoise and the Gulf fishing bat.

The Gulf contains a so-called "disjunct" segment of its fish fauna, which, although not strictly categorized as endemics, are biogeographically and evolutionarily interesting: several species occurring in the northern part of the sea are absent from its southern (more tropical) waters, but appear again in the cooler waters on the outer coast of the Baja California peninsula and extend northward to southern and central California, or even farther. Examples of such disjunctly distributed fishes are: leopard shark, bat ray, giant sea bass, spotted sand bass, California scorpionfish, Mexican rockfish, white sea bass, sargo, pink surfperch, California sheephead, rock wrasse, bay blenny, bluebanded goby, longjaw mudsucker, shadow goby, fantail sole, and diamond turbot, among others. The origins and relationships of this disjunct fauna have been of long-term interest to biologists and have been the subject of some recent studies.

Species at risk: minke, sei, Bryde's, blue, fin, gray, humpback and sperm whales; vaquita (critically endangered); totoaba; Gulf corvina; several species of large groupers (*Mycteroperca spp.* and *Epinephelus spp.*); several species of large snappers (*Lutjanus spp.*); some species of snooks (*Centropomus spp.*); and loggerhead, East Pacific green, leatherback and olive ridley sea turtles. Species such as silky shark, dusky shark, blacktip shark, lemon shark, Pacific sharpnose shark, broadnose sevengill shark, Pacific angel shark, several species of houndsharks (*Mustelus spp.*), scalloped hammerhead and smooth hammerhead sharks, diamond stingray, bat ray, longnose eagle ray, golden cownose ray, and California butterfly ray are presenting signs of overexploitation from artisanal shark and ray fisheries targeted at immature individuals. Other species of concern include whale shark, small- and largetooth sawfish, and manta rays (giant manta, smoothtail mobula, pygmy devil ray, and sicklefin devil ray).

Important introduced and invasive species: a red algae (*Ishige foliacea*), Pacific giant oyster, and threadfin shad. Although there are not yet any documented impacts from introduced species in marine habitats in the Gulf of California, the introduction of domestic animals and pests to the Gulf's islands have eliminated many birds, rodents and lizards from those ecosystems.

Key habitat: coastal lagoons, estuaries, esteros, deltaic systems, mangroves, seagrass beds, coral communities and (few) coral and (many) rocky reefs, rhodolith beds and hydrothermal vents; an important oceanic recruitment area for billfish species is located between Los Cabos and Puerto Vallarta

Human activities and impacts: the Gulf of California contributes to approximately 50 percent of Mexico's national fisheries production by volume; fisheries production has declined over the last decade. Offshore fisheries impact species with low fertility rate such as the pelagic thresher, bigeye thresher, great white shark, and shortfin mako. Recently, mega-resort/tourism/vacation properties developments have commenced, including new marinas for increased recreational watercraft, which are principally aimed at US and Canadian residents and investors, and are rapidly proceeding with little ecological oversight (e.g., Cabo San Lucas/Los Cabos, La Paz, Loreto, San Felipe, Puerto Vallarta/Nuevo Vallarta, Mazatlán, Guaymas/San Carlos, Puerto Peñasco).

---

18 Average at 10 meters depth.
19 This percentage does not include the shelf area around the Midriff Islands' straits.

---

and hydrothermal vents, and supports specialized biotic communities that are based on the use of hydrogen sulfide instead of sunlight for energy. The Gulf of California's shores vary from silty to sandy to rocky, with most of medium-size sands.

Although it is a west coast region, the moderating effect of the Pacific Ocean on the climate is greatly reduced by an almost uninterrupted chain of 1,000 to 3,000-m-high mountains along the Baja California peninsula. The climate of the region is therefore more continental than oceanic, a fact that contributes to the large annual and diurnal temperature ranges observed. Desert-like conditions (Sonoran Desert) are found at its northern end and most of the Baja California peninsula (annual rainfall less than 100 mm). Generally, monsoon rainfall conditions can be observed from the end of June until September in the entire region (for example, annual rainfall in Nayarit can amount to as much as 1,000 mm), and there is more precipitation on the eastern than on the western side of the Gulf of California. Total rainfall depends on the incidence of tropical storms. Hurricanes seasonally affect the lower Gulf and often extend farther northward, and winds are extremely variable. In general, the region exhibits more tropical/subtropical characteristics during summer and more temperate characteristics during winter, especially in its northern part.

As most river water has been impounded or diverted for use in agricultural and urban purposes, freshwater input by rivers is relatively small (Santamaría-del-Angel *et al.* 1994). The Gulf of California is an evaporative basin and exchange with the open Pacific is minor.

The rocky reefs of the Gulf of California provide refuge for a wide variety of territorial and demersal fish. *Photo:* Octavio Aburto

The Gulf has mainly three natural mechanisms that help nourish the region: wind-induced upwelling, tidal mixing, and thermohaline circulation (Alvarez-Borrego 2002). Although a complex pattern, upwelling generally occurs off the eastern coast with northwesterly winds ("winter" conditions from December through May), and off the Baja California coast with southeasterly winds ("summer" conditions from July through October), with June and November as transition periods. After the upwelling events, which usually last only for a few days, the water column stabilizes and phytoplankton communities bloom. Dissipation of tidal energy is strongest in the upper Gulf and around the Midriff Islands. Tidal amplitudes, which may be large as 7 m in the northernmost Gulf, and tidal mixing have a net effect of carrying colder and nutrient-rich water to the surface. Generally, heat and salt are exported out of the Gulf into the open Pacific and, as a result of thermohaline balances, the inflowing deep water has a higher inorganic nutrient concentration than the outflowing surface water. On the other hand, El Niño (ENSO) events have a suppressing effect on the primary productivity of the Gulf. These events can cause reproductive and recruitment failure of organisms higher in the water column and on and around the islands due to

suppression of primary production or changes in the planktonic community structure, or both (Alvarez-Borrego 2002). Wind-driven upwelling is better developed and extends over a greater distance along the east coast than off the Baja California peninsula. Very low oxygen concentrations at intermediate depths (300 to 900 m) are characteristic of the Gulf waters.

Biological Setting

The Gulf of California is mainly a subtropical system (but closer to a temperate system in its northern part during winter) with exceptionally high rates of primary productivity due to a combination of its topography, warm climate, and systems of upwellings. This high primary productivity supports large populations of Pacific sardine, thread herrings, and many species of anchovies (*Anchoa, Anchovia, Cetengraulis, Engraulis*) which are in turn the main food source of a whole array of piscivorous species, including squids, fishes, seabirds, dolphins and whales. The Gulf and its islands also serve as breeding areas for seabirds and marine mammals. For instance, much of the world's population of the widely distributed Heermann's gull, royal tern, brown pelican,

The Spanish shawl gets its color from a carotenoid pigment that it finds in the orange polyps of its only prey, the stickyhydroid. Bay of Los Ángeles, Baja California.
Photo: Octavio Aburto

The whitetip reef shark, a requiem shark, is one of the most important nocturnal predators of the rocky coral reefs. Roca Partida, Revillagigedo Archipelago. *Photo:* Octavio Aburto

long-beaked common dolphin, and California sea lion breed in the region. The region is also home to a whole suite of other species, from the yellow-footed gull to the Mexican rockfish, and from the Pacific seahorse to the blue whale. Waters of the Gulf are so rich as to support a small, largely reproductively isolated population of fin whale, *Balaenoptera physalus*, the year round; an anomalous situation for this (seasonally) highly migratory cetacean elsewhere (Urbán *et al.* 2005).

Overall, the Gulf of California supports a diverse fish fauna composed of 911 species, 821 of them bony fish and 90 of them

cartilaginous fish species (L.T. Findley, pers. comm.). There are also almost 5,000 known macroinvertebrate species in the Gulf, which are estimated to be less than half of the actual biodiversity (Hendrickx *et al.* 2005). Throughout the Gulf, mollusks and crustaceans are the most diverse taxa of macroinvertebrates. Cetacean species diversity in the Gulf of California is very high and its 31 species (in 21 genera), which are present permanently or seasonally, represent 39 percent of world's total cetacean diversity.

Hundreds of species are dependent upon the riparian and aquatic habitats of the Colorado River delta, even though these habitats

The blue-footed booby is an agile diving bird whose breeding range extends from the Gulf of California to Peru.
Photo: Patricio Robles Gil

Blue whales give birth to calves every two to three years. At birth, calves are about
23 feet (7 m) long and weigh 5,000 to 6,000 pounds (2,700 kg).   *Photo:* Patricio Robles Gil

have been deprived of a great deal of their natural freshwater input due the river's impoundment for agricultural and urban purposes. The upper Gulf provides habitat for many marine endemic species such as totoaba, Gulf corvina, and the vaquita—the most endangered cetacean in the world. It also acts as nursery habitat for many species, including shrimp, several species of corvinas and croakers, sharks, rays and others of commercial importance. Other species at risk in the Gulf of California include various species of sharks, rays, groupers, snappers, great whales—including minke, sei, Bryde's, fin, gray, humpback, blue and sperm—and loggerhead, East Pacific green, leatherback and olive ridley sea turtles.

Human Activities and Impacts

Although the Gulf of California is thought to be a resilient large ecosystem, due in part to its coastal wetlands and submarine topography/surface wind patterns that cause the upwelling of nutrients; factors such as overfishing, river water diversions, sedimentation, pollution, and aquaculture installations (mainly shrimp farms) have been altering the region's ecosystems. The decrease of fresh water input from the Colorado River to the Gulf of California has drastically changed the ecological conditions of what used to be an estuarine system, important for fish reproduction. It is now an area of high salinity, and many ecological processes occurring in the once brackish waters have been diminished or altered, including changes in life history patterns of important species such as totoaba (Rowell *et al.* 2008).

Fishing in the Gulf is of prime importance to local communities and to Mexico in general, but current fishing levels have exceeded maximum sustainable levels in most commercial fisheries. Apart from several species of sharks, rays, and mantas (see *Species at risk* above), commercially fished species in the Gulf of California include: blue, brown and whiteleg shrimps; swimming or blue crabs; several species of clams, murexes and pen shells; jumbo squid; northern anchovy; Pacific sardine, round herring, and middling and deepbody thread herrings; sawtail, broomtail, leopard, goliath, and star-studded groupers, Gulf coney, spotted and flag cabrillas, and spotted, goldspotted and parrot sand basses (all from the *Serranidae* Family); amarillo, colorado, spotted rose, Pacific dog, Pacific red, and whipper snappers and barred pargo (Family: *Lutjanidae*); Gulf, shortfin, striped, and orangemouth corvinas, and bigeye croaker (Family: *Sciaenidae*); almaco, green, black, Pacific crevalle and yellowtail jacks, along with blackblotch pompano and jack mackerel (Family: *Carangidae*); roosterfish (Family: *Nematistiidae*); dolphinfish or dorado (*Coryphaenidae*); striped mullet (*Mugilidae*); skipjack, black skipjack, bigeye, yellowfin, and Pacific bluefin tunas, as well as Pacific chub mackerel, wahoo, and Gulf and Pacific sierras (Family: *Scombridae*); sailfish and black, Indo-Pacific blue and striped marlins (*Istiophoridae*); dappled flounder, Cortez halibut and fantail sole (*Paralichthyidae*); finescale and orangeside triggerfishes (*Balistidae*); and bullseye puffer (*Tetraodontidae*). For many years, the use of hook-and-line type gears was able to support a healthy fishery—one that depended on long life spans and decades of egg and larval production by fishes in an ecosystem subjected to relatively little fluctuation and environmental perturbation. As stocks declined in abundance, fishermen have moved to other gear and to newly targeted species. With higher rates of fishing mortality and the escalation of gear types to gillnets, trawls and longlines, a fairly rapid reduction in total standing stocks, changes in species dominance, and the loss of older age classes of larger fish followed. It is believed that the region's stocks will soon decline to levels that will not produce maximum sustainable yield. The totoaba, a large and endemic fish of the northern Gulf of California—for which a legal fishery no longer exists—along with other apex predators of the Gulf of California that appear to have declined to very low levels, and declining stocks of highly migratory species (marlins, sailfish, tunas), may be examples or perhaps harbingers of such a dreadful prospect. Overfishing, the loss of ecologically important species via bycatch mortality, and the widespread destruction of bottom habitats by shrimp trawlers, all contribute to the worsening situation. A formerly large fishery for sea cucumbers collapsed in recent years due to overfishing to supply oriental markets. The continuing excessive mortality of large pelagic predators and the shift in biomass dominance to planktivorous species could have substantial and possibly irreversible effects on the ecological structure and function of the region, triggering a broad expansion of ctenophores (comb jellies), squids (for which a new fishery, again for oriental markets, is developing for one of the larger species recently abundant), and small pelagic fishes (sardines, anchovies, etc.) (NOAA 2002).

The Baja California peninsula, until recently known for its remoteness, is becoming increasingly populated along several Gulf shores, where the coast is otherwise dotted with many fishing villages. Urban development, in general, has not been a major threat to the region. However, during the years 2000–2006, a mega-development project (*Escalera Nautica*) was planned and set into motion. With the planned construction of several new coastal marinas and support installations, and the renovation of existing facilities at some port towns and cities, the project is aimed at luring 1.6 million recreational boat owners to the Gulf of California. This coastal development may increasingly threaten the ecology of the region.

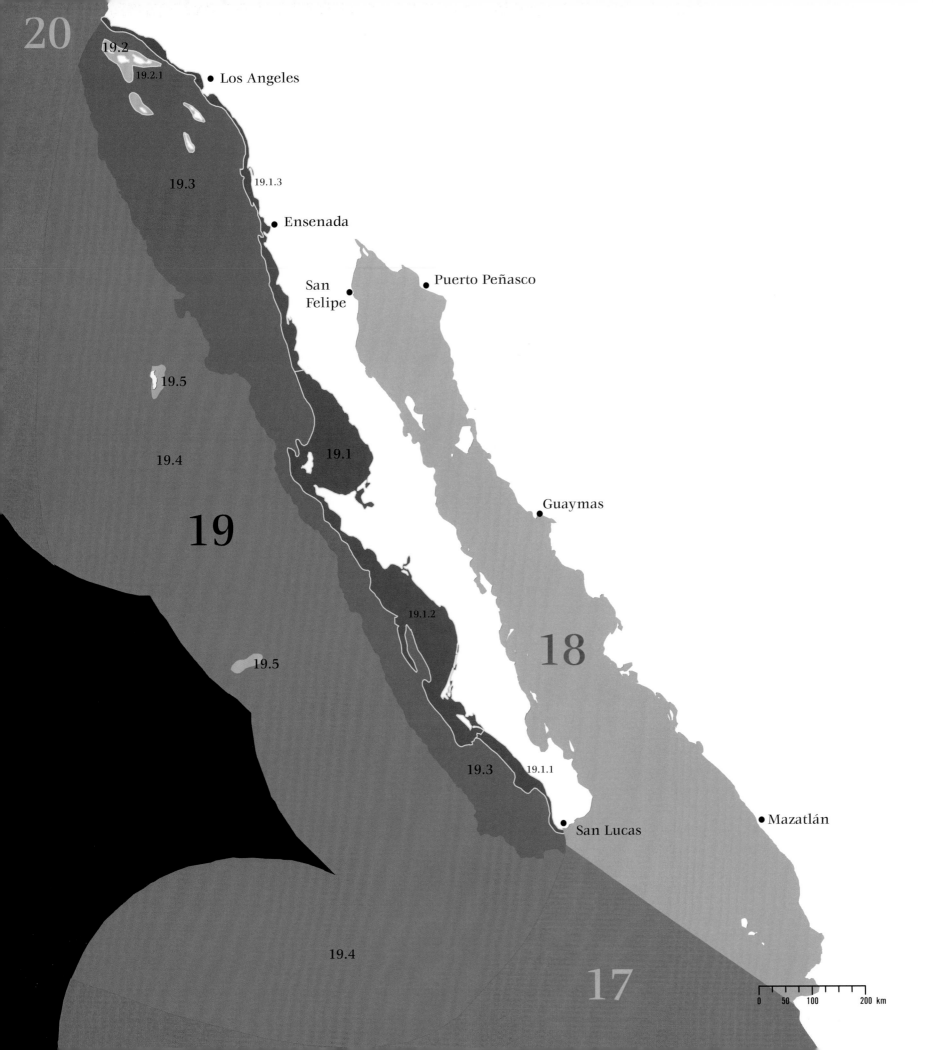

20

19.2

19.2.1

● Los Angeles

19.3

19.1.3

● Ensenada

San
Felipe ●

● Puerto Peñasco

19.5

19.4

19.1

19

19.1.2

18

● Guaymas

19.5

19.3

19.1.1

● Mazatlán

● San Lucas

19.4

17

0 50 100   200 km

19. Southern Californian Pacific

Level II seafloor geomorphological regions include:

Level III coastal regions include:

Regional Overview

The Southern Californian Pacific, which stretches along the Pacific Coast from the southern tip of Mexico's Baja California peninsula at Cabo San Lucas, north to Point Conception, California in the United States, is a region where both waters and faunas mix from north and south. The confluence of the rich, cold, more northerly California Current and warmer waters from the south make this region a complex transition zone between warm and cold temperate biota, characterized by relatively high species diversity. Productivity is moderately high due to the coastal upwelling systems of the region, which bring nutrients to the surface near shore. The region also includes the northern-most extension of mangrove and the southernmost extension of kelp in the eastern Pacific. Abundant stands of giant kelp found around the offshore islands and along the mainland provide a home for a wide variety of invertebrates, fishes, seabirds and marine mammals.

The region includes seven B2B Marine Priority Conservation Areas (PCAs): PCA 17-Upper Bight of the California/Channel Islands/San Nicolas Island (shared with the Montereyan Pacific Transition Ecoregion); PCA 18-Lower Bight of the Californias/ Islas Coronados; PCA 19-Bahía San Quintin/Bahía El Rosario; PCA 20-Isla Guadalupe; PCA 21-Vizcaíno/Isla Cedros; PCA 22-Laguna San Ignacio; and PCA 23-Bahia Magdalena (Morgan *et al.* 2005).[20] But, in contrast, it also includes highly urbanized coastal areas of Southern California (USA) and Tijuana and Ensenada (Mexico).

Physical and Oceanographic Setting

The Southern Californian Pacific is characterized by a very narrow continental shelf, which widens slightly in the south to between 110 and 140 km at Bahía Sebastián Vizcaíno, and just north of Bahía Magdalena. At the shelf break, the seafloor drops rapidly to 1,000 and 3,000 m depths. Seaward of the shelf, but landward of the Southern Californian Plains and Seamounts are the Baja California Borderlands—a mountainous region, found at depths between 800 and 1,000 m, that includes islands, banks and deep basins. The Southern Californian Pacific also includes islands emerging from its abyssal plain, such as Isla Guadalupe, Rocas Alijos and the Channel Islands (which include San Clemente, Santa Catalina, San Nicolas, Santa Cruz, Santa Rosa, and San Miguel). The bottom type found in the coastal regions varies from sandy to rocky.

---

20   See footnote 6 (p.11).

Fact Sheet

Rationale: a region of mixing—of both waters and faunas—from north and south, characterized by relatively high species diversity. The northern boundary is a major biogeographic transition zone for fish and invertebrates, the northern range terminus of many low-latitude species, and the southern range terminus of many high-latitude species.

Surface: 909,679 km²

Sea surface temperature: 15–18°C (winter), 19–22°C (summer)

Major currents and gyres: California Current and Southern California Countercurrent

Other oceanographic features: temperate sea; mixed semidiurnal tides; Baja California Frontal System.

Physiography: geomorphologically complex region

Depth: shelf (roughly 0–200 m): 8%; slope (roughly 200–2,500/3,000 m): 20%; abyssal plain (roughly 3,000+ m): 72%

Substrate type: sand, rock

Major community type and subtype: coastal lagoons (with mangroves in the south), seagrass beds, rocky shore, tidal pools, sand shores, kelp beds, continental platform bottom communities

Productivity: Moderately high (150–300 g C/m²/yr). The region undergoes seasonal upwellings of cold, nutrient-rich water that generate localized areas of high primary productivity that support fisheries for sardines, anchovy, and other pelagic fish species.

Endemics: white, green and pink abalones

Species at risk: minke, sei, Bryde's, blue, gray and fin whales; Guadalupe fur seal; southern sea otter; California least tern; cowcod and boccacio rockfish; white, green, black and pink abalones. Species such as silky shark, scalloped hammerhead, and smooth hammerhead are presenting signs of overexploitation from artisanal shark fisheries targeted at immature individuals. Other species of concern include whale shark, basking shark, great white shark, Garibaldi damselfish, sawfishes and mantas (giant manta, spinetail mobula, smoothtail mobula, pygmy devil ray, and sicklefin devil ray).

Key habitat: breeding and calving lagoons of the gray whale; Ensenada and San Lucas oceanographic fronts are highly productive areas

Human activities and impacts: fisheries, coastal tourism and urban development. Offshore fisheries may be affecting species with low fertility rates such as pelagic thresher and bigeye thresher.

A group of charismatic bottlenose dolphins.   *Photo:* François Gohier

The climate is arid to semi-arid, and has limited freshwater input to the coast. In the northern part, to approximately 350 km south of the US-Mexico border, a Mediterranean climate is found.

The region is affected by various currents and upwellings at different times of year. The region is dominated oceanographically by the north-to-south flowing California Current that carries relatively cold nutrient-rich waters. At Point Conception, the California Current moves farther offshore, allowing the near-shore current to be influenced by the warmer Southern California Countercurrent (which has a discontinuous northward flow beginning seasonally in August-October and strengthening in winter). Southern California Countercurrent and extensions of the Costa Rica Coastal Current influence the coastal region, mainly during winter. Intense coastal upwelling events occur during spring and summer. Seasonal upwelling cells southward of prominent capes and headlands occur near Point Conception, Cape Colonett, Punta Baja, Cape San Quintín, Punta Eugenia, Punta Abreojos, and Cape Falso. Counterclockwise gyre systems occur in the Southern California Bight and Bahía Sebastián Vizcaíno.

The Baja California Frontal System (BCFS) is a dynamic region covering a zone 500 by 250 km, centered about 150 km off the Pacific coast of Baja California Sur. It is characterized by a persistent high concentration of frontal features, generated by the confluence of the cool southbound California Current and warmer northbound Davidson Current as it intersects the Baja California Peninsula. The BCFS appears more active under La Niña conditions (Etnoyer *et al.* 2004).

Biological Setting

The region's confluence of warm southern waters and colder northern ones give it relatively high species diversity. The southern range terminus of many high-latitude marine fishes, invertebrates, and algae as well as the northern range terminus of many equatorial species occurs around Point Conception and the northern Channel Islands (Airamé *et al.* 2003). Productivity in the Southern Californian Pacific is moderately high due to the coastal upwelling systems, which bring nutrients to the surface near shore. With intensive upwelling, also comes greater recruitment success for commercially important fish stocks.

Fish hiding under the graceful and "winged" giant manta ray.
Photo: François Gohier

Laguna Ojo de Liebre, a lagoon located in the Vizcaíno Biosphere Reserve and used by gray whales for mating and giving birth to their calves since time immemorial, was severely exploited during the 19th century by whale hunters. *Photo:* Chris Johnson/EarthOCEAN

Fish migrate large distances to the southern California Bight between upwellings to spawn. El Niño Southern Oscillation (ENSO) conditions, which bring warm equatorial waters further north, decrease productivity and recruitment success of many species, and hinder community dynamics in the region. These interannual variations are superimposed on Pacific Decadal Oscillations, consisting of a sequence of warm and cold regimes. ENSO events have a great impact on this region during a warm regime, reducing the abundance, diversity and stability of the near-shore giant kelp community.

The Southern Californian Pacific includes the northernmost extension of mangrove habitats in the eastern Pacific (close to Bahia Sebastian Vizcaino), and the southernmost extension of giant kelp beds (near Bahia Magdalena). Giant kelp beds at depths of 6 to 30 m are among the most productive marine habitats, providing food and shelter for numerous invertebrates like the Spanish shawl and its preferred prey, the stickyhydroid, as well as fishes, seabirds and marine mammals.

The region also supports large seabird and marine mammal populations. Pink-footed shearwater, short-tailed albatross and Xantus' murrelet—all highly migratory seabirds at risk of extinction—use the productive waters in this region for feeding. The breeding colonies of Xantus' murrelet are found between Islas San Benito and Guadalupe Island and the northern California Channel Islands. Major California sea lion rookeries occur at the Channel Islands and Bahia Sebastian Vizcaino, and important northern elephant seal rookeries occur at San Miguel Island, Santa Barbara Island, Islas San Benito and Guadalupe Island. The Guadalupe fur seal—a transboundary species at risk, with a very limited range—is found in these waters between Isla Guadalupe off the Baja California peninsula and San Nicolas Island off southern California. The species' breeding and pupping occurs on Isla Guadalupe as well as Isla San Benito del Este, Mexico. Likewise, Laguna Guerrero Negro, Laguna Ojo de Liebre (Scammon's Lagoon), Laguna San Ignacio, Santo Domingo Channel, and Bahia Magdalena are the most important breeding and calving areas for the gray whale—a species with one of the longest

Adult Garibaldi damselfish in an algae "forest." Santa Barbara, California.
Photo: Steven Wolper/DRK PHOTO

migratory routes of all mammals (22,000 km yearly from the Bering Sea to Baja California). Over 27 species of whales and dolphins also visit the Channel Islands (USA).

Human Activities and Impacts

The Southern Californian Pacific Region includes highly urbanized coastal areas of southern California and Tijuana, as well as sparsely populated coasts of Baja California and Baja California Sur. Los Angeles, Orange and San Diego counties, bordering the Southern California Bight, have the first-, fifth- and sixth-largest populations in the United States, respectively, collectively totaling more than 15 million people. Despite rapid increases in population, wastewater discharges of most pollutants into US waters have been decreased by 50 to 99 percent since the 1970s, resulting in improvements in benthic/demersal and kelp communities, and reductions in contaminants in fish and marine mammals.

The region is rich in a variety of fishery resources. Anchovies and sardines are key links in the local trophic system. Other commercial fishes include Pacific chub mackerel, Pacific bonito, jack mackerel, Pacific hake and over 60 species of rockfishes. In the US Pacific coast (ecoregions 19, 20 and 21), six out of 48 federally managed stocks are overfished, with 13 of unknown status (NMFS 2007).

Coastal regions vary significantly in the degree of human alteration—from relatively uninhabited to highly modified—and include major centers of marine transportation, recreation and offshore oil production. For the once relatively pristine but rapidly developing area of the northern Baja California peninsula coast, and its neighboring southern California coast with its thriving economy, issues of concern stem from oil and gas development, busy shipping lanes, non-point sources of pollution, riverbed exploitation for sand, thermoelectrical plants, vacation homes and tourism infrastructure, as well as commercial and recreational fishing.

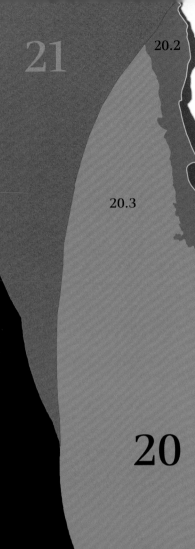

21

20.2

20.1

20.3

20.1.1

20.1.2 ● San Francisco

20

20.2

20.1

● Los Angeles

● Ensenada

San
Felipe ● ● Puerto Peñasco

18

20.3

19

Guaymas ●

0   50   100   200 km

20. Montereyan Pacific Transition

Level II seafloor geomorphological regions include:

20.1   Montereyan Pacific Transition Shelf
20.2   Montereyan Pacific Transition Slope and Canyon System
20.3   Montereyan Pacific Transition Plain and Seamounts

Level III coastal regions include:

20.1.1   San Francisco Bay Estuarine Area
20.1.2   Montereyan Neritic

Regional Overview

The Montereyan Pacific Transition Region, which stretches along the central California coast from Point Conception to Cape Mendocino (USA),[21] has moderately high productivity associated with the seasonal upwelling that occurs along its coasts. Also found in the region is a series of very significant submarine canyons and seamounts, such as the Monterey Submarine Canyon—one of the largest on the Pacific coast of North America. Because of the canyon's proximity to shore, deepwater species of whales, dolphins and seabirds are found near the coast. The region also includes three major estuaries that serve as important habitats for many marine species. San Francisco Bay—the largest estuary in the region—is a major staging area for migratory birds in the Pacific Flyway, but is also an area where more than 95 percent of the historic tidal marshes have been modified. Invasive species are also a major threat to the Bay's biota. The sea otter, a keystone species, also makes its home within the region. The region includes two B2B Marine Priority Conservation Areas (PCAs): PCA 16-Central California, and PCA 17-Upper Bight of the Californias/Channel Islands/San Nicolas Island (Morgan *et al.* 2005).[22]

Physical and Oceanographic Setting

The Montereyan Pacific Transition Region consists of a very narrow continental shelf and steep continental slope, transected by a series of submarine canyons. Three major estuaries are found in northern and central California, within San Francisco Bay and Tomales Bay, and Elkhorn Slough. Shelf habitats are predominantly composed of soft sediments, although rocky areas are the most biodiverse. The deepest and largest submarine canyon on the coast of North America is the Monterey Canyon in the center of Monterey Bay. It is 470 km long, approximately 12 km wide at its widest point, has a maximum rim to floor relief of 1700 m, and serves as a major conduit for sediment transport from the continental shelf to the deep-ocean floor. Other major canyons include Bodega, Pioneer, Carmel, Sur and Lucia Canyons. Beyond the slope, Gumdrop, Pioneer, Guide, and Davidson Seamounts rise above the abyssal plain.

---

21   The northern boundary between this region and the Columbian region is approximate, and other authors have identified ecoregional boundaries in the vicinity of Monterey Bay (California, USA), a primary biogeographic break for marine algae, shallow-water benthic invertebrates and fishes (Airamé *et al.* 2003); Point Arena (California, USA - e.g., NERRS 2004, http://nerrs.noaa.gov/bioregions/coverage.html); Cape Mendocino (California, USA – e.g., current system, Hayden et al. 1984, Kellerher *et al.* 1995), or Cape Blanco (Oregon, USA – e.g., Strub *et al.* 2002, based on physical factors).
22   See footnote 6 (p. 11).

Fact Sheet

Rationale: transitional between temperate and subtropical regions and faunas

Surface: 337,281 km²

Sea surface temperature: 11–14°C (winter), 13–15°C (summer)

Major currents and gyres: California Current, California Undercurrent, Davidson Current and Southern California Countercurrent

Other oceanographic features: strong upwelling in spring

Physiography: Very narrow shelf with major canyon systems in the slope and below.

Depth: shelf (roughly 0–200 m): 4%; slope (roughly 200–2,500/3,000 m): 13%; abyssal plain (roughly 3,000+ m): 83%

Substrate type: sand and rock from Point Conception to Estero Bay; mud-sand and rock north to Monterey Bay; mud-sand sediments in San Francisco Bay; and mainly sandy north of San Francisco Bay

Major community type and subtype: bays and estuaries, sandy beach and rocky intertidal communities, kelp beds, submarine canyon and cold seeps, deep sea and seamount communities, offshore island and bank communities

Productivity: Moderately high (150–300 g C/m²/yr). The effects of coastal upwelling, ENSO and the Pacific Decadal Oscillation result in strong, inter-annual variability in productivity.

Endemics: delta smelt and subpopulations of several Pacific salmon species

Species at risk: blue, fin, North Pacific right, humpback, gray and sperm whales; Steller sea lion; southern sea otter; California brown pelican; California least tern; marbled murrelet; leatherback, loggerhead; delta smelt; steelhead, chinook and coho salmon; cowcod and bocaccio rockfish; black and pinto abalone

Important introduced and invasive species: over 234 species identified (affecting San Francisco Bay, in particular), including Asian clam, compound sea squirt, Chinese mitten crab, and European green crab

Human activities and impacts: tourism, fishing, commercial shipping, coastal development

The California Current is the dominant current system affecting the region. However, the sub-surface, poleward California Undercurrent flows northward along the continental slope, extending to the surface next to the inshore Davidson Current from October through February. During periods of low stratification in the winter, the Southern California Countercurrent, the Davidson Current and the California Undercurrent can merge, and this seasonal northward flow helps account for this region's transitional nature. Current fluctuations occur in association with El Niño events and basin-wide events (e.g., Pacific Interdecadal Oscillation). Within the California Current system, upwelling is most pronounced in this region, occurring off major capes, particularly between mid-February and mid-July.

Biological Setting

The region represents a transitional zone between the subtropical species representative of southern California and Baja California, and the more northerly species. The major biogeographic affinities seem to be with the northern regions, but southern species often extend their ranges during ENSO events and warm phases of the Pacific Decadal Oscillation. The strong seasonal upwelling contributes to moderately high productivity.

The brown pelican is a social and low-flying bird that prefers areas near the coast.   *Photo:* Patricio Robles Gil

A school of blue rockfish swirls under the canopy of a kelp bed forest. *Photo:* Norbert Wu/Minden Pictures

This remarkably productive coastal environment is home to numerous mammals, seabirds, fishes, invertebrates and plants, such as giant kelp, krill, dungeness crab, rockfish, bonito, California skate, Pacific salmon, market squid, albatross, shearwaters, common murre, ashy storm-petrel, brown pelican, gulls, Steller sea lion, Dall's porpoise, harbor seal, as well as gray, blue and humpback whales.

Coastal wetlands associated with estuaries support millions of shorebirds and waterfowl during spring and fall migrations, and over the winter months. These estuaries also serve as important spawning and nursery grounds for marine species. San Francisco Bay is the largest estuary in the region and is a major staging area for migratory birds in the Pacific Flyway, hosting roughly a million migratory and resident birds.

The region's kelp beds are key habitats for numerous species. Giant kelp forms dense beds on rocky subtidal areas and bull kelp, which is the most abundant surface canopy kelp in California north of Santa Cruz, occurs from near Point Conception, California, north to the eastern Aleutian Islands. Within this region, the southern sea otter is resident in areas between Point Conception and San Francisco Bay. The sea otter is regarded as a keystone species because of its significant influence in maintaining kelp forest communities, primarily through its predation on sea urchins, the dominant herbivore. Major northern elephant seal areas are found south of Cape Mendocino, with a large rookery at Ano Nuevo Point. Davidson Seamount is one of the largest seamounts on the west coast and has remarkable biological communities, including large, dense patches of sponges and extremely old gorgonian coral aggregations, with individuals commonly reaching heights of more than 3 m.

Human Activities and Impacts

The Montereyan Pacific Transition Region includes scenic coastlines and the San Francisco urban area—the second-largest urban population on the west coast of North America. The San Francisco Bay estuary is renowned for its natural beauty, international commerce, recreation, and sportfishing. However, more than 95 percent of the historic tidal marshes have been modified, with attendant losses in fish and wildlife habitat. The flow of freshwater into the estuary has been greatly reduced by water diversions largely to support irrigated agriculture. Harbor and channel dredging, including both activities undertaken during dredging for navigation and disposal of dredge spoils, disturb communities, alter water flow patterns and salinity. Contaminants also enter the estuary through municipal and industrial sewage, as well as urban and agricultural runoff. Phosphorus concentrations are high and sediment and fish contaminant conditions are poor (EPA 2005). Invasive introduced species are a major threat to the Bay's biota.

Fisheries in the region are important, but have suffered major declines. Numerous species, including rockfishes, sardines, salmon, sablefish, and abalones, have declined under pressure of commercial and recreational fishing. Some species, including lingcod, cowcod, bocaccio, canary rockfish, and Pacific ocean perch are overfished, and salmon and steelhead populations have been listed under the US Endangered Species Act.

In the US Pacific coast that includes ecoregions 19, 20 and 21, six out of 48 federally managed stocks are overfished and 13 are of unknown status (NMFS 2007).

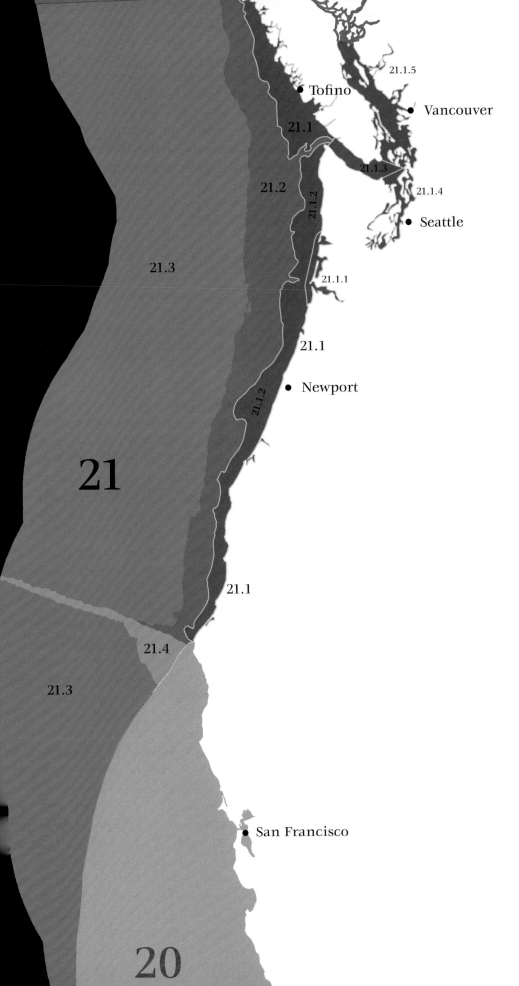

21.1.5

Tofino

Vancouver

21.1

21.1.3

21.1.4

21.2

21.1.2

Seattle

21.3

21.1.1

21.1

21.1.2

Newport

21

21.1

21.4

21.3

20

San Francisco

0   50   100   200 km

21. Columbian Pacific

Level II seafloor geomorphological regions include:

Level III coastal regions include:

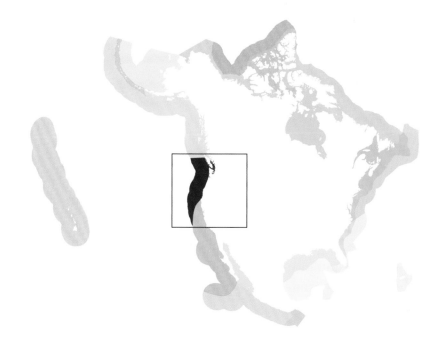

Regional Overview

The Columbian Pacific Region, home to both the Columbia and Fraser Rivers, has high runoff with vast amounts of nutrients that stimulate growth of phytoplankton, algae and other marine life. The region also has seasonal upwelling that contributes to moderately high productivity. The Columbian Pacific is home to the greatest oyster and clam production in North America as well as major adult concentrations of killer whale. Forestry, fishing, shipping, tourism and marine recreation are the main activities that contribute to the region's high standard of living. With these lucrative activities, however, comes ecological stress, notably in the Salish Sea area. The region stretches along the Pacific coast from Cape Mendocino in the south, to the Strait of Juan de Fuca, the Strait of Georgia, Puget Sound, and north on the seaward side of Vancouver Island, to Cape Cook. The Columbian Pacific Region includes three B2B Marine Priority Conservation Areas (PCAs): PCA 13-Southern Strait of Georgia/ San Juan Islands; PCA 14-Barkley Sound/Pacific Coastal Washington; and PCA 15-Central Oregon/Cape Mendocino (Morgan *et al.* 2005).[23]

---

23 See footnote 6 (pg 11).

Physical and Oceanographic Setting

The Columbian Pacific is characterized by a moderately narrow continental shelf and steep continental slope. The seafloor of the Juan de Fuca plate, north of the Mendocino escarpment that runs west from Cape Mendocino, is shallower than the seafloor of the Pacific plate to the south. Offshore seafloor features include the Heceta Banks, which rises 100 m above the edge of the continental shelf to within 80 m of the ocean surface. Its mass and depth cause the California Current to flow over or around it, introducing eddies and other instabilities that affect areas far downstream and along the Oregon coast. Underwater canyons at the edge of the continental shelf, such as the Astoria Canyon, Rogue River Canyon, and Juan de Fuca Canyon, have unique habitats and set up their own upwelling conditions that concentrate nutrients into areas of high topographic relief, thus driving high levels of biologic productivity.

The coast includes extensive tracts of forests, dunes, estuarine areas, and rocky shorelines with spectacular coastal headlands and tidal pools, and off Vancouver Island, many islands, large bays and sounds, as well as fjordal inlets. The region has higher runoff than the regions to the south, especially from the

Fact Sheet

Rationale: the region has a temperate fauna and flora, quite different than its northern and southern neighbors

Surface: 574,781 km²

Sea surface temperature: 9–11°C (winter), 13–15°C (summer) relatively warm surface waters in inland sea locations

Major currents and gyres: California Current, Davidson Current, Vancouver Island Coastal Current

Other oceanographic features: Juan de Fuca Eddy is of high productivity

Physiography: mountainous shoreline with a relatively narrow shelf, widening at the Heceta escarpment; the Straits of Juan de Fuca, Georgia and Puget Sound are semi-enclosed bodies with estuarine influences; complex ridges, canyons and channels are found in deeper waters

Depth: shelf (roughly 0–200 m): 10%; slope (roughly 200–2,500/3,000 m): 12%; abyssal plain (roughly 3,000+ m): 78%

Substrate type: mostly sand with areas of rock nearshore and rock, gravel and mud-sand offshore

Major community types and subtypes: bays and estuaries, sandy beach and rocky intertidal communities, kelp forests, benthic and pelagic communities of the continental shelf, submarine canyons and cold seeps, deep sea and seamounts, offshore islands and banks

Productivity: moderately high (150–300 g C/m²/yr)

Endemics: none known, although there are some salmon endemic subpopulations

Species at risk: blue, fin, north Pacific right, humpback, killer, gray and sperm whales; sea otter; marbled murrelet; leatherback sea turtle; Pacific salmon (chinook, coho and chum); steelhead; Pacific hake, cowcod rockfish, bocaccio; black and pinto abalones

Important introduced and invasive species: Over 100 invasive species have been identified in estuaries. Important species include Japanese eelgrass, Atlantic cordgrass, purple varnish clam, Asian clam, sea squirt, and European green crab.

Human activities and impacts: forestry, fishing, shipping, tourism and marine recreation

Columbia and Fraser Rivers. The former drains a watershed of approximately 671,000 km². In the spring and summer, Columbia's plume creates a surface lens of lower-salinity water spreading west and south in the California Current as far south as Cape Mendocino. The Fraser River's plume carries vast amounts of nutrients to the ocean of the North East Pacific, but it flows northwards year around off Vancouver Island as the Vancouver Island Coastal Current, before dissipating at about the northern end of Vancouver Island. Both freshwater discharges stimulate growth of phytoplankton, algae and other marine plant life. The Strait of Georgia and Puget Sound are distinguished by their mostly rocky and gravel shores with discrete extensive mud flats, and by the pronounced summer and fall thermal stratification of the water column. Surface waters frequently exceed 20°C in the summer and fall.

The California Current is the dominant summer and offshore current system affecting the region while the wind-driven Davidson Current is the dominant nearshore winter current system. Upwelling occurs seasonally off major capes from February through September, and near the southern end of Vancouver Island, encouraging a prolific ocean ecosystem. A poleward undercurrent flows continuously along the shelf break from 33°N to 51°N at an average depth of 200 m. The Strait of Juan de Fuca has an estuarine circulation, with outflowing surface waters and deep, inflowing oceanic water.

Biological Setting

The Columbian Pacific Region has a temperate fauna and flora, lacking many of the subtropical species that intrude into its southern neighbor, the Montereyan Pacific Transition Region. The Strait of Juan de Fuca represents a major faunal discontinuity. The seasonal upwelling of the region contributes to moderately high productivity. The Strait of Georgia and Puget Sound, collectively often referred to as the Salish Sea, are the largest and most important estuaries of the region, forming unique coastal environments. Willapa Bay is the next-largest of the estuaries. Together, these areas are the location of greatest oyster and clam production in North America.

The numerous rocks, bays and islands provide valuable habitat for many seabirds, such as the common murre, as well as marine mammals, including the Steller and California sea lions. Submerged rocky reefs are also scattered along the coast. These areas and their associated kelp beds provide valuable habitat for a wide variety of marine species. Bull kelp is one of the most abundant surface canopy kelp north of Santa Cruz. Within kelp beds, the sea otter—one of the smallest marine mammals—plays an important role in structuring the nearshore coastal ecosystem and for this reason is often held up as an example of a marine keystone species.

Major adult concentrations of killer whale begin north of the Columbia River. The region is also host to gray, blue, minke, and humpback whales. Some other common species of the region include harbor seals, scoters, rockfish, Pacific herring and all species of northeast Pacific salmon. Populations of pinnipeds and cetaceans have grown since the 1970s, in part in response to increased protection under the American Marine Mammal Protection Act of 1972.

Human Activities and Impacts

Throughout the region, forestry, fishing, shipping, tourism and marine recreation are the main human activities contributing to the high standard of living prevailing in the region. These lucrative and popular activities, along with pollution from ship traffic, urban runoff, destruction of shoreline habitats, and industrial pollution combine, however, to create the main sources of ecological stress for the region, notably in the Salish Sea area. Developments

Racing upriver en masse to spawn, pink salmon are one of many species that rely on oceanic and freshwater river habitats to survive. *Photo:* Michael Quinton/Minden Pictures

The tall dorsal fin and whitish patches are signature features of the killer whale, especially common in the waters off the Pacific Northwest. Salmon are a preferred prey.   *Photo:* Flip Nicklin/Minden Pictures

The ochre sea star (*Pisaster ochraceus*) is a keystone species for rocky intertidal communities from Baja California to Prince William Sound, Alaska. The ochre star is predominantly purple but in other areas, a bright orange and almost yellow phase is also found.
Photo: Darrell G.Gulin/DRK PHOTO

in major estuaries and deltas have altered and reduced important habitats, while unsustainable fishing pressures failures have had serious impacts on a variety of fish and shellfish populations and other organisms that depend on them. Watershed development and freshwater diversion have altered the estuaries throughout the region, and may have dramatically affected marine production. Some of the most recent activities being developed in the region are salmon and shellfish aquaculture. In this region, as in the Atlantic, the potential ecological impacts of these industries are still being debated, with some indication that they may have negative effects on natural marine ecosystems. Some cultured species are intentionally introduced exotics to the region, and are included in the more than 100 marine exotic species that are now found in the Strait of Georgia alone. However, by far and away the greatest source for the introduction of these invasive species is purged ballast water from shipping.

The Georgia Basin/Puget Sound region is currently home to nearly six million people—double the population of the mid-1960s. Most estuarine wetland habitat has been modified or destroyed since 1850, and much of the shoreline has been altered by human development through port facilities, piers, and other shoreline armoring structures, as well as through dredging, filling or diking. Many smaller estuaries, however, remain in a natural state and provide important biological services. Much of the rest of the Pacific coast in this region is relatively sparsely populated.

Fisheries have traditionally been a mainstay of the region's economy. Pacific salmon, many rockfish and lingcod have shown significant population decreases. The dramatic decrease in traditional salmon harvests and heavy fishing pressures on other groundfish species have resulted in a shift toward near-shore reef fisheries for recreational fisheries, while commercial groundfish fisheries are searching deeper and further offshore. The live-fish fishery and the sport bottomfish fishery focus their efforts in this rocky reef habitat, and the effects these fisheries have on fish populations have not been fully assessed. Populations of Pacific herring and other fishery species have greatly decreased since 1975. This decrease is thought to be due to changing predation by harbor seals and Pacific hake, alteration of nearshore habitat, especially in eelgrass beds, and possibly changing water conditions such as varying temperatures. Populations of many species seem to be greatly influenced by large-scale ocean climatic regime shifts in the North Pacific as a whole, and these changes are just beginning to be factored into analyses of fish stock status.

The Columbia River generates electric power for residents and businesses, provides irrigation for crops, and harbors deepwater ships that come and go across the Pacific. Millions of people depend on this and the Fraser River for employment in water-related industries, for commerce, and for transportation. These rivers, in particular, are under mounting pressure from cumulative impacts, including stormwater runoff, industrial discharges, fishing, development, irrigation, power generation, forestry, mining, transportation, and water removal—all which affect the estuarine areas, anadromous fishes, and nearby marine ecosystems. In the coastal US Pacific Northwest that includes ecoregions 19, 20 and 21, six out of 48 federally managed stocks are overfished, and 13 are of unknown status (NMFS 2007).

The pink-footed shearwater is found well offshore in the open ocean waters over the continental shelf. *Photo:* Mike Danzenbaker

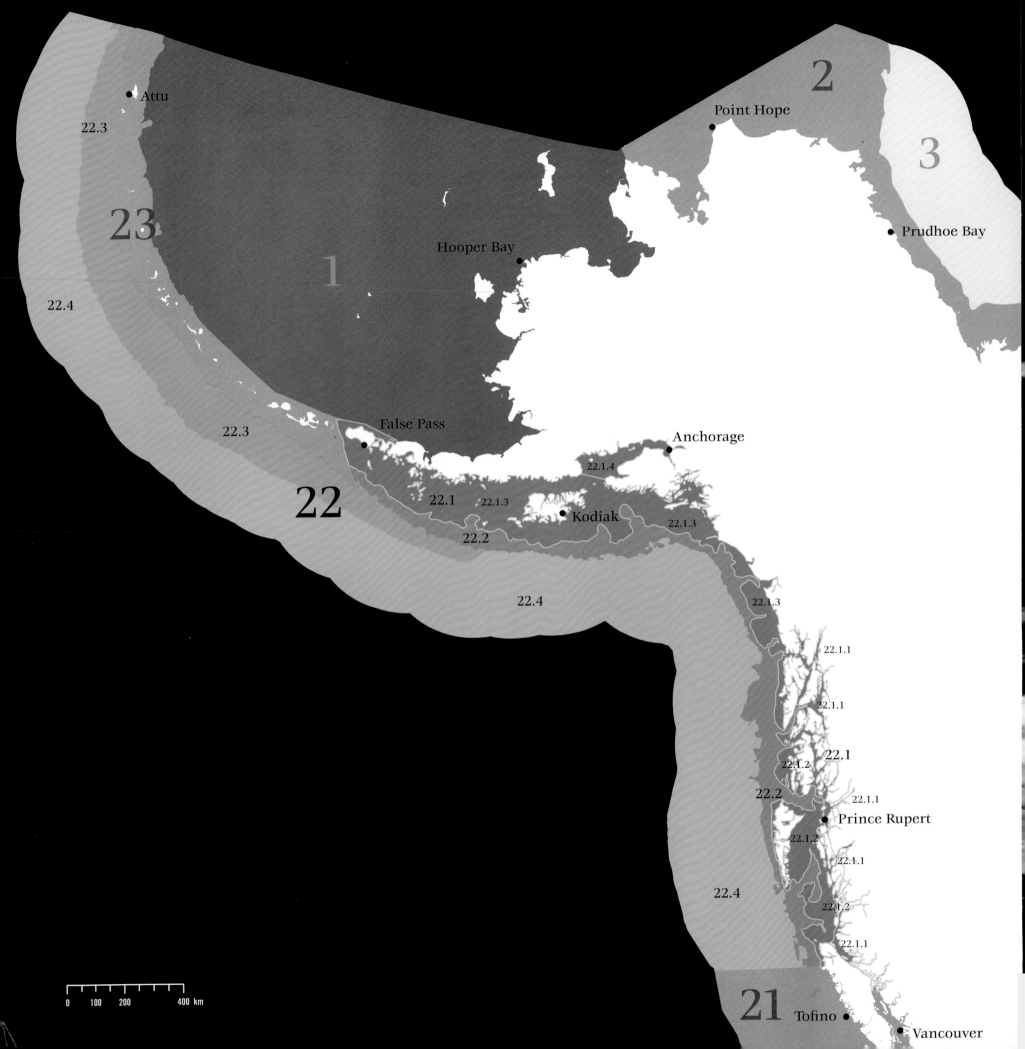

22. Alaskan/Fjordland Pacific

Level II seafloor geomorphological regions include:

22.1   Alaskan/Fjordland Shelf
22.2   North Pacific Slope
22.3   Aleutian Trench
22.4   North Pacific Basin

Level III coastal regions include:

22.1.1   Fjordland Estuarine Areas
22.1.2   Fjordland Neritic
22.1.3   Gulf of Alaska
22.1.4   Cook Inlet

Regional Overview

The Alaskan/Fjordland Pacific Region is a binational region, home to abundant plant and wildlife. The region encompasses a multitude of islands, deep fjords, sheltered straits and many of the world's gazetted seamounts. Its adjacent rivers carry a tremendous amount of freshwater runoff and nutrients, while upwelling in the center of the Alaska Gyre pushes nutrients, phytoplankton and zooplankton production onto its shelf. The region straddles Vancouver Island, starting at Cape Cook on the west side and the Strait of Georgia on the east. It continues north through the Gulf of Alaska and extends to the end of the Aleutian Archipelago Region, running south and west of that region. The region includes eight B2B Marine Priority Conservation Areas (PCAs): PCA 5-Western Kodiak Island/Shelikof Strait; PCA 6-Lower Cook Inlet/Eastern Kodiak Island; PCA 7-Prince William Sound/Copper River Delta; PCA 8-Patton Seamounts; PCA 9-Glacier Bay/Sitka Sound/Fredrick Sound; PCA 10-Dixon Entrance/Langara Island/Forrester Island; PCA 11-Northern Queen Charlotte Sound/Hecate Strait/Gwaii Haanas; and PCA 12-Scott Islands/Queen Charlotte Strait (Morgan *et al.* 2005).[24]

---

24   See footnote 6 (p.11).

Physical and Oceanographic Setting

The Alaskan/Fjordland Pacific Region encompasses all of the fjord-dominated west coast of British Columbia and the Alaska Panhandle, and runs out to sea over the North Pacific Slope and Basin. Its numerous islands, deep fjords, and sheltered straits, as well as the great amount of freshwater runoff from its numerous rivers, distinguish the southern portion of the region. The shelf of the Fjordland varies, but is generally narrow throughout—it extends about 20 km at the north end of Vancouver Island, is almost imperceptible at the southern end of Queen Charlotte Islands, and broadens again as it moves north into the Gulf of Alaska, where it is about 160 km wide. The deeper waters of the region include over five percent of the world's gazetted seamounts.

The major oceanographic influence on the region is the Alaska Current, which is formed when the westerly North Pacific Current (West Wind Drift) bifurcates at Vancouver Island, splitting into the northerly, counterclockwise-directed Alaska Current and the southerly California Current. The Alaska Current continues along the southern edge of the Aleutian Islands as the Alaska Stream, and a parallel, low-salinity Alaska Coastal Current flows close to the coast from British Colombia to Unimak Pass. Based

Fact Sheet

Rationale: separated from the Columbian Region by the bifurcation of the North Pacific Current to form the cooler Alaska Current

Surface: 2,029,679 km²

Sea surface temperature: 1–9°C (winter) and 10–16°C (summer) and reaching 20°C in sheltered areas during the warmest months

Major currents and gyres: Alaskan Current/Stream, Alaska Coastal Current, North Pacific Current

Physiography: rocky coastlines, numerous islands, fjords and embayments, narrow continental shelf, large number of seamounts rise from the deeper waters offshore

Depth: shelf (roughly 0–200 m): 18%; slope (roughly 200–2,500/3,000 m): 22%; abyssal plain (roughly 3,000+ m): 60%

Substrate type: mainly rock and mud inshore with sand, rock and gravel offshore

Major community types and subtypes: mud flats, tidal marshes, rocky reefs, rocky shorelines, kelp beds, eelgrass beds, seamounts, hydrothermal vents

Productivity: moderately high (150–300 g C/m²/yr)

Endemics: none known; however, seamount fauna have not been well studied

Species at risk: bowhead, sperm, sei, beluga, right, humpback, gray, and blue whales; Steller sea lion; sea otter

Important introduced and invasive species: at least seventeen non-indigenous species identified in south-central Alaska, including Norway rat, Arctic fox, Atlantic salmon, European green crab, purple loosestrife, and Japanese knotweed

Human activities and impacts: fishing, marine recreation, tourism, oil and gas exploration and recovery

on its stable temperatures, the Alaskan/Fjordland Region may be considered a transition zone between the polar Bering Sea and Arctic Ocean, and the temperate waters of the mid-latitude Pacific Ocean.

Sea ice is generally absent from this region. The land barrier imposed by the Alaskan Peninsula prevents much of the cold Arctic currents from flowing down the west coast, so there is little oceanic water exchanged between the Arctic and lower Pacific regions. Ice occurs only seasonally at the northern boundary near the Bering Sea, the Sea of Okhotsk, and in northern bays and inlets where glaciers may feed into the ocean.

Biological Setting

The Alaskan/Fjordland Shelf hosts one of the most highly productive marine ecosystems in the northern Pacific. Upwelling in the center of the Alaska Gyre pushes nutrients, phytoplankton and zooplankton production onto the shelf along the edge of the gulf. The region is home to about 3,800 species of invertebrates, representing 3.5 percent of all marine invertebrates in the world. These populations include a rich mixture of oceanic, subpolar, neritic (living in the tide waters and landwashes) and benthic plankton.

The large invertebrate populations provide rich food sources for the 306 species of fishes living in the region. The Pacific herring is the most abundant, while steelhead, dolly varden, Pacific cod, walleye pollock, Pacific halibut, arrowtooth flounder, and five species of salmon—coho, chinook, chum, pink and sockeye—are also abundant. Shellfishes such as clam, crab, scallop, shrimp, and squid are common. Over the years, salmon and herring stocks have been heavily fished, and although herring stocks are rebounding, the health of salmon stocks remains precarious. Climate and current changes in the 1970s associated with the Pacific Decadal Oscillation caused population shifts in commercially important fish and shellfish species and subsequent declines in populations of sea lions and puffins. The Kodiak Archipelago is home to the Kodiak bear, the region's terrestrial apex predator and a species that has been genetically isolated from other bear populations for some 12,000 years.

The region is important for a large proportion of the world's populations of Cassin's auklet (70 percent—particularly on the Scott Islands, which includes 55 percent of the world population), ancient murrelet (40 percent), as well as some 75 percent of Canada's tufted puffins. The region also provides feeding and resting areas for large numbers of migrating and wintering ducks, geese (*Anser* and *Branta spp.*), swans, loons, and shorebirds.

Marine mammals common to the region are gray, minke, humpback and killer whales; harbor and Dall's porpoise; Pacific white-sided dolphin; and sea otter. Although the sea otter's range is vast, the greater part of the world's population can be found in Alaskan waters. Major concentrations of adult sea otter can be found within this region in the Alexander Archipelago, in Prince William Sound, and on Kodiak Island. The North Pacific right whale has major adult concentrations off Kodiak Island and the Alaska Penninsula, and major adult feeding concentrations of humpback whales occur from Kodiak Island to Unimak Pass, in Prince William Sound and the Alexander Archipelago. The Cook Inlet stock of beluga whale declined by nearly 50 percent between 1994 and 1998 and has been listed as depleted under the US Marine Mammal Protection Act. The northern fur seal is the most abundant seal in the region, and harbor seals are also common. Steller sea lions, which have major rookery areas in the Alexander Archipelago and on the Alaska Peninsula, have declined since the 1970s.

Through the continental portion of the region (and adjacent areas), freshwater discharges from the Fraser, Skeena, Nass, Stikine, Susitna and other rivers carry vast amounts of nutrients to the ocean, stimulating the growth of phytoplankton, algae, and other marine plant life. Along the water's edge, coastal salt marshes and mud flats contain large beds of eelgrass, important spawning sites for Pacific herring schools and nursery areas for some salmon. In the subtidal zones lie vast forests of giant kelp and bull kelp. Recent explortion of seamounts in the Gulf of Alaska and Aleutian Islands have revealed rich faunas of deepwater stony corals and gorgonians (especially of the Families *Paragorgiidae, Primnoidae*)—unique deepwater ecosystems.

Human Activities and Impacts

The region ranges from the coastal urban areas of just north of southwest British Columbia (with one of the fastest growing human populations in North America), to sparsely inhabited areas further north. Throughout the region, fishing, shipping, tourism and marine recreation are the main human activities. But with these lucrative and popular activities have come pollution from ship traffic, urban runoff, degradation of shoreline and bottom habitat, heavy fishing, and industrial pollution—the main sources of ecological stress to the region. Development in major estuaries and deltas has altered and reduced habitats, while unsustainable fishing pressures have had serious impacts on a variety of fish and shellfish populations and other organisms that depend on them. One of the latest activities being developed is aquaculture—for finfish (salmon) and shellfish (mussels, oysters, scallops), in particular—some of which rely on introduced species. The spread of diseases and parasites is also a concern associated with aquaculture. In this region, as in the Atlantic, the potential ecological impacts of the industry are still being assessed and debated as there are concerns that it might have negative effects on marine ecosystems. Ecological conditions of the coastal resources in Alaska are poorly known. Alaska has assessed less than 0.1 percent of its coastal estuaries (EPA 2005). While most coastal areas are relatively unspoiled, there are pockets of contamination. In the US northern Pacific and western ecoregions (1, 2, 22 and 23), two out of 35 federally fished stocks are overfished (NMFS 2007).

Kachemak Bay State Park, Alaska, at low tide
Photo: Patricio Robles Gil

Kodiak bears need kilometers of land free of the presence of humans to forage. This is why two-thirds of Kodiak Island (Alaska) is designated a national wildlife refuge.
Photo: Patricio Robles Gil

Humpback whales feeding on herring, which they have "corralled" using a bubble curtain. *Photo:* Brandon D. Cole

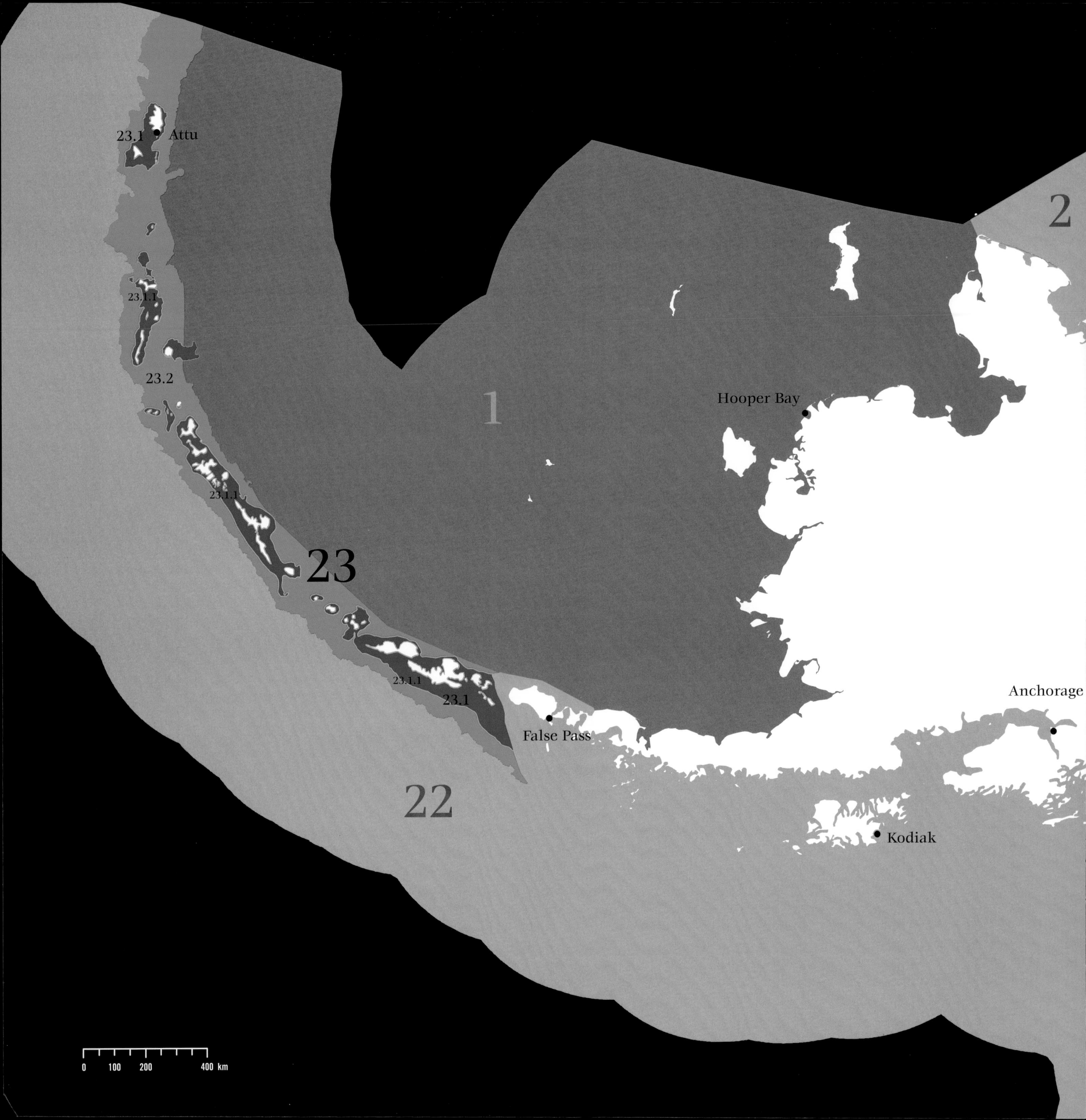

23.1 • Attu

23.1.1

23.2

23.1.1

23

1

2

Hooper Bay

Anchorage

23.1.1

23.1

False Pass

Kodiak

22

0   100   200   400 km

23. Aleutian Archipelago[25]

Level II seafloor geomorphological regions include:

23.1   Aleutian Shelf
23.2   Aleutian Slope

Level III coastal regions include:

23.1.1   Aleutian Neritic

Regional Overview

The Aleutian Archipelago Ecoregion houses the world's longest archipelago and is adjacent to the Aleutian Trench—a deep-sea feature 3,700 km long and 7,680 m deep. Numerous high-velocity straits and passes connect the temperate North Pacific to the subpolar Bering Sea along the Archipelago, with the greatest flow being northward from the lower latitude Pacific Ocean to the Arctic. The region is considered a transition zone between the polar seas of the Bering and the Arctic and the temperate waters of the mid-latitude, northern Pacific Ocean. The region is also home to major concentrations of adult sea otters—however, in recent years populations have declined dramatically. In addition, amongst the wide variety of deep-sea coral species found in the region, large colonies of red-tree coral, up to 500 years old, have also been discovered here. The unique combination of rich nutrients and underwater volcanoes has created diverse and abundant coral habitat. Although the Archipelago is largely uninhabited, many human activities affect the region, such as fishing, shipping, tourism and marine recreation. The Region includes two B2B Marine Priority Conservation Areas (PCAs): PCA 3-Western Aleutian Islands/Bowers Bank; and PCA 4-Unimak Pass/Aleutian Islands (Morgan *et al.* 2005).[26]

Physical and Oceanographic Setting

West of the Gulf of Alaska, the Aleutian Islands—the world's longest archipelago—extend westward towards Russia. This region is characterized by a narrow shelf, little freshwater input and no seasonal ice cover. The shelf slopes steeply offshore to the Aleutian Trench. The Aleutian Trench runs for 3,700 km from Kodiak Island to the end of the Aleutian Chain, and has a maximum depth of 7,680 m. The Aleutian Trench extends in an arc south of the Aleutian Islands, where the Pacific plate slides under the North American plate. This subduction zone is along the Ring of Fire, the string of volcanoes and frequent-earthquake zones around the Pacific Ocean.

The Aleutian Archipelago contains numerous straits and passes, with swift currents, connecting the temperate North Pacific to the subpolar Bering Sea. The Alaskan Stream flows westward out of the Gulf of Alaska along the southern edge of the Aleutian Islands. As this flow continues along the Aleutian Islands, the majority of shallower water enters the Bering Sea through Near

25 Information from the following section was adapted from *Marine Priority Conservation Areas: Baja California to the Bering Sea*, written for the CEC and MCBI by Lance Morgan, Sara Maxwell Fan Tsao, Tara A.C.Wilkinson and Peter Etnoyer (2005).
26 See footnote 6 (p. 11).

Fact Sheet

Rationale: long archipelago system with a narrow shelf, little freshwater input and strong influence of the Alaska Current/Stream

Surface: 180,620 km²

Sea surface temperature: 1–10°C

Major currents and gyres: Alaska Current/Stream and Aleutian North Slope Current

Physiography: narrow shelf that drops off on the Pacific side to the deep Aleutian Trench

Depth: shelf (roughly 0–200 m): 23%; slope (roughly 200–2,500/3,000 m): 74%; abyssal plain (roughly 3,000+ m): 0%

Major community types and subtypes: gorgonian coral and sponge gardens

Productivity: Moderately high (150–300 g C/m²/yr), affected by large-scale atmospheric and oceano-graphic conditions

Species at risk: bowhead, sperm, right, blue, humpback and sei whales; Steller sea lion; sea otter; short-tailed albatross

Important introduced and invasive species: Several introduced species, including rats and foxes, threaten colonies of seabirds

Key habitats: major nesting and feeding habitat for seabirds; rich deep-sea coral and sponge habitats

Human activities and impacts: fishing, oil and gas exploration and recovery

Strait, strongly influencing water properties and circulation in the eastern Bering Sea. The mean circulation through the Aleutian passes, the eastern Bering Sea and the Bering Strait is northward. Thus, little polar oceanic water is exchanged between the Arctic and the lower-latitude Pacific Ocean. Nutrient-rich water is introduced to shallow zones—where it can be used by phytoplankton—via strong mixing within the Aleutian passes. The Aleutian Archipelago may be considered a transition zone between the polar seas of the Bering and Arctic, and the temperate waters of the mid-latitude, northern Pacific Ocean.

Biological Setting

Rocky shores throughout the Aleutians contain abundant bull kelp forests. These systems are also home to major adult concentrations of sea otter, although these populations have been in steep decline in recent years. Exploration of seamounts in the Gulf of Alaska and Aleutian Islands has revealed unique deepwater ecosystems of rich faunas associated with deepwater hydrocorals and gorgonians, especially the northern octocoral, *Paragorgia arborea,* and also the red-tree coral, which provides structural habitat for the rockfish, sablefish, Atka mackerel and arrowtooth flounder found in the region. Large colonies of red-tree coral may be 500 years old. Colorful and beautiful gorgonians extend to depths of 730 m and appear in aggregations like groves of trees. Some of these corals rise more than 4.5 m above the sea floor. There are at least 44 known species of deep-sea corals in Alaska and species diversity may well rival tropical coral reefs.

The unique combination of rich nutrients and underwater volcanoes has created diverse and abundant coral habitat.

Leatherback sea turtles are occasionally sighted as far north and west as the Aleutian Islands. Nearly 40 million seabirds, representing 30 species, breed among these islands. Near the center of the Aleutian Archipelago there are several locations crucial for birds, including key feeding habitat (Atka Pass), a large fulmar colony (Chagulak Island) and a notable feeding habitat for whiskered auklets (Sitkin Sound and Islands of Four Mountains). The world's largest colony of crested and least auklets is on Kiska Island and Alaska's largest colony of tufted puffins nests on Kaligagan Island. Additionally, the Canada goose breeds in the region and overwinters in Mexican wetlands, and both Laysan albatross and pigeon guillemot cross into Canadian and lower US waters to feed.

North Pacific right whales concentrate in the Aleutian Islands and major adult feeding concentrations of humpback whales occur from Kodiak Island to Unimak Pass. Blue whales feed near the Aleutian Islands and Bering Sea before heading south to southern California and Mexico to breed and calve. Northern elephant seals feed in the region before returning south to the coasts of California, Pacific Mexico and Isla Guadalupe to breed and molt. In the Aleutians, there are 10,000 Steller sea lions. Although they have major rookery areas in the Aleutian Islands, their numbers have declined by 75 percent since the 1970s (Angliss and Lodge 2002).

Fisheries in the Aleutians include a larger number of species, including walleye pollock, Atka mackerel, rockfishes, sablefish, Pacific cod, arrowtooth flounder, Pacific halibut, Greenland turbot and others. Salmon (sockeye, Chinook, pink, Coho and chum), during their spawning runs, constitute a major food source for Kodiak bears.

Human Activities and Impacts

The Aleutian Archipelago consists mostly of uninhabited islands; however, many kinds of human activities have affected the region. Throughout the region, fishing is the predominant human influence, although shipping; tourism and marine recreation are all increasing. Since 1989, Dutch Harbor/Unalaska has averaged in excess of 226,800 tons (500 million pounds) of annual commercial fish landings, and is consistently the port with the largest annual landings in the United States (National Marine Fisheries Service (NMFS) fishery statistics[27]). The trawl fishery for walleye pollock and Atka mackerel is the main threat to the region's biodiversity.

---

27 NOAA Fisheries Annual Commercial Landing Statistics, http://www.st.nmfs.gov/st1/commercial/landings/annual_landings.html.

Showing its down-turned yellowish tufts on the back of the neck, a tufted puffin swims away.   *Photo:* Patricio Robles Gil

Regardless of the apparent high concentrations in the rookeries, there is increased concern about the status of northern fur seal populations, particularly in the Aleutian Islands, where there has been a roughly 50% decrease in pup-production since the 1970s.
Photo: Stephen J. Krasemann/DRK PHOTO

A common murre splashing on the water.   *Photo:* Patricio Robles Gil

Most dramatic is the decline of most fish-eating seabirds, probably as a result of commercial fishing which has caused a drastic drop in available food, as well as entanglement of the seabirds in fishing gear. Bottom trawling and longlining are threats to the deep-sea coral beds found throughout the Aleutians. In 2005, the North Pacific Fishery Council recommended banning bottom trawling from major portions of the Aleutian Archipelago, in part to protect rich coral and sponge habitats. Restrictions on trawling around Steller sea lions colonies are presently in place. Moreover, historical overhunting of great whales is likely to have influenced the trophic dynamics of the region. In the US northern Pacific and western Arctic ecoregions (1, 2, 22 and 23), two out of 35 federally fished stocks are overfished (NMFS 2007).

Alien species also take their toll on these relatively unpopulated islands. Several introduced species, including rats and foxes, threaten colonies of seabirds. Pollution too is severe in certain areas, mostly from active and inactive military bases. Nuclear testing on Amchitka Island in 1971 resulted in radioactive isotopes entering the ecosystem.

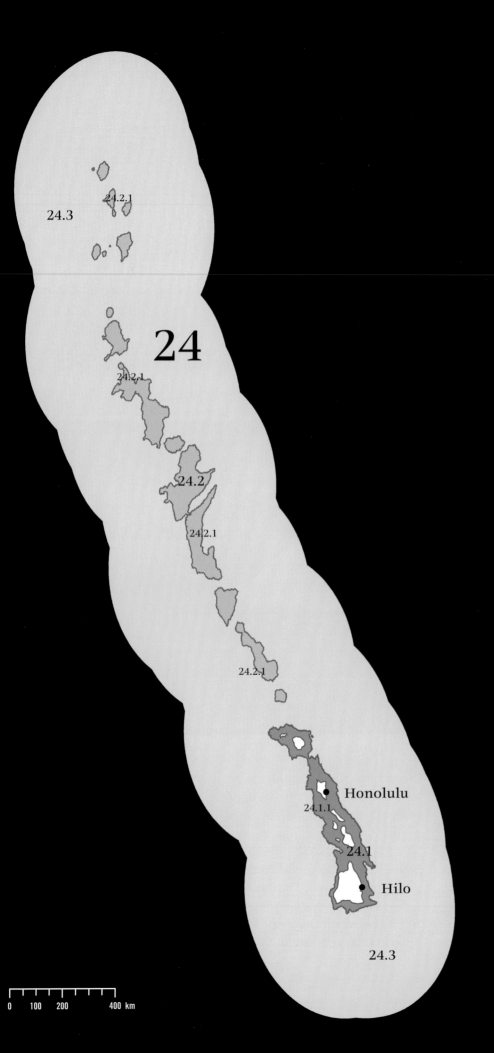

24

24.3

24.2.1

24.2.1

24.2

24.2.1

24.2.1

24.1.1

Honolulu

24.1

Hilo

24.3

0   100   200   400 km

24. Hawaiian Archipelago

Level II seafloor geomorphological regions include:

24.1 Main Hawaiian Islands, Reefs and Banks
24.2 Northwestern Hawaiian Islands, Banks and Seamounts
24.3 Hawaiian Islands Abyssal Plain

Level III coastal regions include:

24.1.1 Main Hawaiian Islands Coastal
24.2.1 Northwestern Hawaiian Coastal

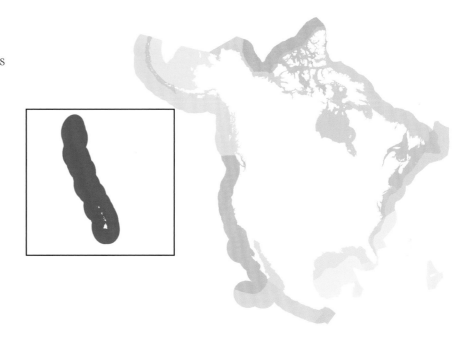

Regional Overview

The Hawaiian Archipelago, one of the most geographically isolated island systems in the world, is composed of eight main volcanic oceanic islands, 124 smaller islands, atolls, banks, and numerous seamounts. Due to this isolation, the reef fauna of the Archipelago are less diverse than other reefs, especially those elsewhere in the Indo-West Pacific Ocean. This relatively low faunal diversity, however, contrasts with unusually high endemism among the reef species. In addition, the uninhabited Northwest Hawaiian Islands are some of the most pristine coral reef systems worldwide. They are home to the endangered Hawaiian monk seal, some of the largest and most important seabird colonies in the world, and North America's largest nesting ground for the green sea turtle. The region is also a major breeding, calving and nursery area for the humpback whale. On the inhabited Main Hawaiian Islands, many watersheds and nearshore areas have been significantly modified. Their nearshore fish stocks are thought to have declined 80 percent in the last century, and because of their high endemism, invasive species are of particular concern. Despite impacts from ocean outfalls, urban growth and coastal development, coral ecosystems in these islands range from fair to excellent condition and water quality in most areas is good (Friedlander *et al.* 2005). The Hawaiian Archipelago stretches 2,450 km from the Big Island of Hawaii northwest to Kure Atoll.

Physical and Oceanographic Setting

The Hawaiian Archipelago consists of a string of volcanic islands formed by the northwest movement of the Pacific tectonic plate over the stationary Hawaiian "hot-spot." The eight Main Hawaiian Islands in the southwest range in age from one to seven million years and are characterized by high volcanic mountains and narrow fringing reefs. The older Northwestern Hawaiian Islands have undergone submergence with only Necker, Nihoa and Gardner Pinnacles still containing exposed volcanic material. The rest of the chain is composed of coral atolls, small sand islands and submerged banks. The chain continues as the submerged Emperor Seamounts northwest of Kure Atoll, the northernmost atoll in the world. The northern extent of the North Equatorial Current is the major oceanic current affecting the islands, branching along the Hawaiian ridge into a North Hawaiian Ridge Current and gyres in the lee of the islands.

Fact Sheet

Rationale: isolated oceanic island archipelago in the North-Central Pacific with a high degree of endemism

Surface: 2,479,560 km²

Sea surface temperature: 24°C (winter), 23°C (summer) and 27°C (Oahu)

Major currents and gyres: North Equatorial Current, Kuroshiro Current

Depth: Main Hawaiian Islands shelf (roughly 0–200 m): 2%; Northwestern Hawaiian Islands, Banks and Seamounts (roughly 0–200 m): 4%; slope and abyssal plain (roughly 200+ m): 94%

Substrate type: volcanic basalt rock with carbonate reefs

Major community types and subtypes: fringing coral reefs, atolls, coral banks, seamounts, and open-ocean pelagic systems

Productivity: Low (<150 g C/m²/yr).

Endemics: Levels of endemism in the coral reef fish, invertebrate and algal biota in Hawaii are among the highest in the Pacific. Around 25 percent of nearshore fishes and stony corals are endemic. Hawaiian monk seal is the only endemic marine mammal.

Species at risk: Hawaiian monk seal; Hawaiian dark-rumped petrel; band-rumped storm-petrel; Newell's shearwater; short-tailed albatross; green, hawksbill and leatherback sea turtles; humpback whale; Hawaiian goby; the inarticulated brachiopod and a rare endemic Hawaiian rice coral, *Montipora dilatata*

Important invasive and introduced species: Numerous nonindigenous marine invasive species have been identified in the Hawaiian Islands, including 287 invertebrates, 20 algae, and 20 fishes. Invasive marine algae (*Gracilaria salicornia, Hypnea musciformis*) are causing significant economic damage, and the snowflake coral is threatening black corals harvested for jewelry. Mangroves do not occur naturally in Hawaii but introduced species, the red and oriental mangroves, have become invasive. Several introduced marine fish species have become established.

Key habitat: coral reefs

Human activities and impacts: tourism, coastal development, coastal recreation, recreational and commercial fisheries, shipping and military operations

Biological Setting

The isolation of the Hawaiian Islands has resulted in unusually high endemism among the reef fauna. Over 25 percent of Hawaiian reef animals are endemic to the region. In addition, although the clear oceanic waters are low in primary productivity, they do provide the environmental conditions needed for rather rich nearshore coral reef ecosystems. There are approximately 60 species of stony corals, more than 100 sponge species, 1071 marine mollusks species, 884 crustacean species, and 557 species of reef and shore fishes in the Archipelago. The Hawaiian Islands Region is also a major breeding, calving and nursery area for the humpback whale.

The uninhabited Northwestern Hawaiian Islands include habitats and species absent in the Main Hawaiian Islands. Because of their isolation and the limited fishing effort upon them, average fish biomass is more than 260 percent greater than in the Main Hawaiian Islands (Friedlander and DeMartini 2002). Apex predators such as jacks and sharks predominate—composing 54 percent of biomass in the Northwest compared to only 3 percent in the Main Hawaiian Islands. Over 90 percent of the threatened green sea turtles in Hawaii were born at French Frigate Shoals, found in the Northwestern Hawaiian Islands. These islands are also home to the major remaining breeding populations of the endangered Hawaiian monk seal. The deepwater Southeast Hancock Seamount had a unique and productive fish fauna that was overfished in the early 1970s and has yet to recover.

Mother humpback whale escorting her calf in the Pacific Ocean.
Photo: Doug Perrine/DRK PHOTO

The Hawaiian monk seal keeps cool on hot, windless days by lying on damp sand. *Photo:* Frans Lanting/Minden Pictures

Human Activities and Impacts

The Main Hawaiian Islands support a population of 1.2 million people, over 70 percent of which reside on the island of Oahu. Hawaii's nearshore reefs annually contribute nearly $1 billion in gross revenues to the state's economy. In addition to the resident population, nearly seven million tourists visit the islands each year, making tourism the largest industry, much of it dependent upon the marine ecosystems. Anthropogenic stresses of greatest concern on the region's reefs are coastal development and runoff, pollution, recreational overuse, damage from ship groundings, alien species, overfishing, damaging fishing practices, and over-harvesting of ornamental reef species. Many watersheds and nearshore areas in the Main Hawaiian Islands have been significantly modified. Approximately 57 percent of Hawaii's estuarine area is impaired by some form of pollution or habitat degradation, whereas only 2 percent of its coastal shoreline is impaired (EPA

2005). Because of the high endemism of the islands, there are special concerns about the introduction of invasive alien species. The unique patch reef habitats of the Archipelago's largest protected embayment, Kaneohe Bay, have been overgrown by two introduced marine algae species. Moreover, the region's nearshore fishery stocks are thought to have declined 80 percent in the 1900s. The commercially most important fishery species are pelagic tunas, but the reefs provide diverse, culturally and recreationally important fisheries. The Northwestern Hawaiian Islands are mostly uninhabited and fishing and other resource extraction is currently limited to a small bottomfish fishery. Here, the greatest anthropogenic impact is from derelict fishing gear, mostly discarded trawl nets. These islands, atolls and banks are protected in the Northwestern Hawaiian Islands Coral Reef Ecosystem Reserve, the Hawaiian and Pacific Isles Wildlife Refuge as well as in the Midway Atoll National Wildlife Refuge.

Greater amberjack corral a school of bigeye scad. *Photo:* Doug Perrine/DRK Photo

Appendix: US Pacific Island Territories

Several oceanic islands in the Pacific are part of United States territories but have not included in the previous descriptions of the 24 ecoregions. With the exception of Johnston Island, they represent distinct ecoregions. Brief regional overviews are given below.

Johnston Island

Johnston Atoll (16°45'N, 169°31'W) is a small, isolated unincorporated territory of the United States in the Central Pacific. It lies 800 km south of its closest neighbor, French Frigate Shoals (Northwestern Hawaiian Islands) and shares biogeographic affinities with the Hawaiian Archipelago, with evidence of larval transport between the two. Because of these faunal affinities and as both the Hawaiian Archipelago and Johnston Island occur in the oceanic North Pacific Transition Zone Province (Longhurst 1998) they can be considered part of the same ecoregion. Due to its isolation, Johnston Island is relatively species-poor and appears to be recruitment-limited. Since World War II it has been used primarily for military operations.

Northern Line Islands

The US Islands of Palmyra and Kingman Reef are part of the Northern Line Islands. The other islands in this archipelago belong to the Republic of Kiribati. Palmyra and Kingman sporadically bathed by eastward-flowing North Equatorial Countercurrent originating in the high biodiversity regions of the Western Pacific. The atolls have diverse shallow-water habitats and among the highest coral species richness of any Central Pacific island or atoll. Palmyra and Kingman Reef are US Fish and Wildlife Refuges.

Jarvis and Phoenix Islands

Howland and Baker Islands are uninhabited US possessions in the Phoenix Island Archipelago. Although the westward-flowing South Equatorial Current is the primary surface current, the eastward flowing Equatorial Undercurrent interacts with the steep topography of these islands to drive strong, topographically-induced equatorial upwelling, providing nutrient enrichment and high local productivity and biomass—especially of planktivores. The species diversity is much lower than for the Northern Line Islands, reflecting both lower habitat diversity and primary currents from the species-poor eastern Pacific. Jarvis Island is considered part of the Southern Line Islands, but its biogeographic and oceanographic affinities are similar to Howland and Baker Islands, lying as it does in the path of the westward-flowing South Equatorial Current. All of these islands are US Fish and Wildlife Refuges.

Wake Island

Wake Island is an oceanic atoll in the North Pacific Ocean, about two-thirds of the way from Hawaii to the Northern Mariana Islands. It lies at the northern end of the Marshall Islands archipelago in the North Pacific Tropical Gyre Province (Longhurst 1998). Major shallow water habitats are coral reefs and the islands support a number of seabirds, including seasonally breeding albatross. Since World War II, Wake Island has been used primarily for military operations and as an emergency landing location for transpacific flights.

Mariana Islands

The Mariana Island Region includes all the volcanic and raised limestone islands and submerged banks of the Mariana Archipelago (4,115 km west southwest of Hawaii) out to the limits of the EEZ of the US and stretches from Guam Island 825 km north to Farallon de Pajaros (about 550 km south of Iwo Jima, the southernmost of the Ogasawara islands of Japan). The arc of the Mariana Archipelago is flanked by the Mariana Trench, which includes the deepest water on earth (11,034 m) in its southern end near Guam.

Samoan Archipelago

The Samoan Archipelago includes five volcanic islands (Tutuila, Aunu'u, Ofu, Olosega, Ta'u) and two remote atolls (Rose, Swains) within the EEZ of the US encompassing American Samoa, 14° south of the equator. The region also includes the islands in the neighboring independent country of (western) Samoa (e.g., Savai`i and `Upolu) and in the French protectorate of Wallis and Futuna.

Supporting Information

Acronyms and Abbreviations

B2B	Baja California to Bering Sea (as an area or initiative)
CEC	Commission for Environmental Cooperation
CICESE	*Centro de Investigación Científica y de Educación Superior de Ensenada* (Center for Scientific Research and Higher Education of Ensenada) (Mexico)
CWS	Canadian Wildlife Service
DDT	dichlorodiphenyltrichloroethane
DFO	Department of Fisheries and Oceans (Canada)
EEZ	exclusive economic zone
IUCN	International Union for the Conservation of Nature, presently called World Conservation Union
INE	*Instituto Nacional de Ecología, Semarnat* (National Institute of Ecology) (Mexico)
MPA	marine protected area
MSCCC	Marine Species of Common Conservation Concern
NMFS	National Marine Fisheries Service, presently called NOAA Fisheries of the Department of Commerce (US)
NOAA	National Oceanic and Atmospheric Administration (US)
PCA	Priority Conservation Area
PCB	polychlorinated biphenyls
PRBO	Conservation Science (formerly Point Reyes Bird Observatory) (US)
Semarnat	*Secretaría de Medio Ambiente, Recursos Naturales* (Secretariat of Environment and Natural Resources) (Mexico)
UNAM	*Universidad Nacional Autónoma de México* (National Autonomous University of Mexico)
UNEP	United Nations Environment Program
US	United States of America
USFWS	United States Fish and Wildlife Service (US Department of Interior)
PSU	Practical Salinity Unit

Glossary of Common Terms

Abiotic — A non-living component of the environment.

Algal bloom — An abundance of algae along the surface of a body of water.

Anadromous — Fish that ascend rivers from the sea for breeding.

Apex predator — A predatory species that is at the top of the food chain in its ecosystem.

Arctic Oscillation — Opposing atmospheric pressure patterns in the northern middle and high latitudes: in its positive phase, high pressure at mid-latitudes drives ocean storms northward, bringing wetter weather to Alaska, Scotland and Scandinavia, and keeps frigid winter air from penetrating as far into the middle of North America, giving the Great Plains and Great Lakes regions warmer winters. In the negative phase, high pressure lies over the polar region and weather patterns are generally reversed from those of the positive phase.

Atoll — A ring or horseshoe shaped coral reef or string of coral islands that usually encloses a shallow lagoon. It is formed when an oceanic island or seamount, ringed by a barrier reef, sinks or is eroded below sea level, leaving only the coral ring and interior lagoon.

Bank — A broad shallow water region, usually sandy, surrounded by deep water. It is usually associated with high levels of productivity.

Barrier island — A long narrow ridge of sand that runs parallel to the coast but is separated from it by a bay or lagoon. Created by wave and current action, its contours are constantly changing via erosion and accretion. Barrier islands buffer the coast from waves and storms.

Barrier reef — A coral reef that runs parallel to but is far from the shore and is separated from it by a lagoon (or channel) of considerable depth and width. In some cases, the lagoon may be several kilometers wide.

Bathymetry — The depth of water and the ocean floor relative to sea level. Bathymetric charts are the submerged equivalent of an above-water topographic map.

Benthic — Defining a habitat or organism found on a freshwater or marine bottom (compare with *Pelagic*).

Benthos — The collection of organisms (plants and animals) living on or closely associated with the bottom of a body of water, especially the ocean.

Bight — Wide bay formed by a curve in a shoreline.

Bioaccumulation — The concentration of long-lived compounds in the flesh and organs of organisms that ingest prey that have ingested those compounds themselves.

Biodiversity (biological diversity) — The variety of organisms considered at all levels, from genetic variants belonging to the same species through arrays of species to arrays of genera; families, and still higher taxonomic levels; includes the variety of ecosystems, which comprise both communities of organisms within particular habitats and the physical conditions under which they live.

Bycatch — Fish and other marine organisms that are captured incidentally in a fishery, which are not sold or kept for personal use.

Carrying capacity (ecosystem) — The number of living organisms that an ecosystem can support indefinitely while maintaining its productivity, adaptability, and capacity for renewal.

Catch — The amount of a living harvested marine resource.

Cenote — A sinkhole or natural depression in the ground due to the removal of soil or bedrock by water, and in many cases the collapse of the roof of a cavern or subterranean passage. Typically found in the Yucatán Peninsula, Florida and some nearby Caribbean islands where limestone is the predominant subterranean stratum, cenotes often link to vast systems of underground passages and subterranean water bodies, with connections to other cenotes and sometimes the sea. These natural wells have long been the only perennial source of water in much of the Yucatán Peninsula and in Pre-Columbian times they played an important role in Mayan ceremonial rites.

Coastal — The area of the ocean that extends from the seashore to the outer extent of the continental shelf.

Community — A collection of organisms of different species that co-occur in the same habitat or region and that interact through trophic and spatial relationships.

Conservation — The sustainable use as well as protection, maintenance, rehabilitation, restoration, recovery and enhancement of ecosystems, natural habitats and viable populations of species in their natural surroundings.

Continental shelf — A broad expanse of ocean bottom, associated with the submerged edge of continental plates that slopes gently seaward from the shoreline to the shelf break (usually 100 to 200 m depth).

Convergence zone — The line where two oceanic water masses meet, resulting in the sinking of the denser mass.

Dead zone — A marine area having hypoxia—lacking oxygen—virtually devoid of life (as, for example, the dead zone currently the size of New Jersey located in the northern Gulf of Mexico, extending from the Mississippi River delta); associated with eutrophication.

Demersal	Part of the ocean or lake that comprises the water column that is near to (and is significantly affected by) the seabed and the benthos. The demersal zone is just above the benthic zone. The term can also refer to all species that live on, or in close proximity to, the seabed. Fishing gear that works near the seabed also receives the term demersal.	**Estero**	Used to denote non-mangrove coastal lagoons lacking regular freshwater inputs (such as those that occur north of the Midriff Islands in the Gulf of California). They are saltier at their head than at their mouth due to high evaporation and lack of freshwater input (also see "negative estuaries" under "estuary").
Ecoregion	An *ecoregion* is a general term used to describe an ecosystem unit or area. An *ecoregion* is a large area of land or water that contains a geographically distinct assemblage of natural communities that (a) share a large majority of their species and ecological dynamics, (b) share similar environmental conditions, and (c) interact ecologically in ways that are critical for their long-term persistence. It may range from global size units through to site or local level systems.	**Estuarine**	The circulation of fresh and brackish water in estuaries, or, in oceans, of the layering behavior of water of lesser and greater salt contents. Many patterns are possible depending on the volume of fresher water on top and the denser, saltier water on the bottom.
Ecosystem	A dynamic complex of organisms (including humans), and the physical environment (soils/bottom type, water, geology etc.) interacting as a functional unit. They may: • vary greatly in size and composition and display functional relationships within and between systems; • be relatively pristine through to extensively altered by human activities/uses; • be aquatic or terrestrial; • be barren or highly productive.	**Estuary**	A semi-enclosed coastal body of water that has a free connection with the open sea and within which the seawater is usually measurably diluted by freshwater derived from land drainage, such as the tidal mouth of a river where fresh water meets the salt water of a sea or ocean. A positive estuary refers to a situation where freshwater input (river flow and precipitation) exceed losses due to evaporation; whereas a negative estuary refers to a situation where evaporation exceeds freshwater input.
Ecosystem services	Processes by which the environment produces resources such as clean water, habitat for fisheries, and pollination—often taken for granted by many.	**Eurythermic**	Able to survive a wide range of temperatures.
El Niño (El Niño-Southern Oscillation, ENSO)	A disruption of the ocean-atmosphere system in the tropical Pacific having important consequences for weather around the globe. Under normal atmospheric and hydrographic conditions, the trade winds result in upwelling of water along the western coasts. Under El Niño conditions, offshore winds are weaker, upwelling ceases and warm waters extends to the coast. This decreases productivity and recruitment success of many species, and hinders community dynamics of the region.	**Eutrophic/ Eutrophication**	A body of water with a high supply of nutrients and high productivity. The nutrient load is often excessive and derived from anthropogenic sources and the system is usually characterized by algal blooms and, at times, hypoxia (compare with *Oligotrophic*; see also *Dead Zone*).
		Exclusive Economic Zone (EEZ)	A zone 200 nautical miles (370 km) wide along the coast where nations have exclusive rights to any resource. It was initiated by the United Nations Convention on the Law of the Sea.
		Exotic species	A species that is not native to an area and has been introduced intentionally or unintentionally by humans; not all exotics become successfully established. (also see *Invasive*)
Endemic	A species or race native to a particular place and found only there. By contrast, see indigenous.	**Extinction**	The termination of any lineage of organisms, from subspecies to species and higher taxonomic categories from genera to phyla. Extinction can be *local*, in which one or more populations of a species or other unit vanish but others survive elsewhere (also see *Extirpated*), or *global*, in which all the populations vanish.
Endemism	The ecological state of being unique to a particular place, due to physical, biological, and climatic factors.		
Environment	Can be closely linked to the term *ecosystem*. Some, however, construe the term to be exclusive of man, such as in the expressions "man and the environment," "man and the biosphere," "economy and the environment."	**Extirpated**	Status of a species or population that has completely vanished from a given area but which continues to exist in some other location (also see *Extinction, local*).
		Flatfish	Fish whose body is strongly compressed and that has both eyes on one side of the head, such as halibut, flounder, sole and plaice.
		Freshwater	In the strictest sense, water that has less than 0.5% of salt concentration; here, it refers to rivers, streams, creeks, springs, and lakes.
		Fringing reef	A type of reef that forms borders along the shore (at times with a very narrow channel between it and the shore), which acts as an extension of the coastline.
		Frontal system/ frontal zone/front	Areas where major currents converge.

Global distillation (or "grasshopper") effect — The large-scale geochemical process by which persistent organic pollutants and other chemicals evaporate or are volatilized in the warmer, southerly or mid-latitudes and are transported to colder, northerly latitudes by the prevailing atmospheric currents where they condense, settle to earth, and often enter the food web.

Groundfish — A general term referring to fish that live on or near the sea floor. Groundfish are also called bottom fish or demersal fish. Examples include cod, haddock, pollock or flounder.

Gulf Stream — A powerful, warm, northerly current in the Atlantic Ocean that originates in the Gulf of Mexico and then, passing the southern tip of Florida, moves along the eastern seaboard of the United States, Nova Scotia, and Newfoundland before arcing across the Atlantic toward Great Britain. At about 30°W, 40°N it splits, the northern part continuing on to northern Europe and giving that continent a milder climate than it would otherwise have for its latitude, and the southern part recirculating off the west African coast.

Habitat — A functional area used by organism(s) as a life supporting system. Habitat can vary greatly in size and composition. A habitat consists of biotic and abiotic features. At times, it can be closely linked to ecosystems.

Hydrothermal vents — Also known as black smokers, hydrothermal vents are deepwater vents that occur along midocean ridges where plates join, and that emit jets of hot water loaded with nutrients and bacteria. Biotic communities of hydrothermal vents are based on hydrogen sulfide instead of sunlight for energy.

Hypersalinity — The state of estuaries when freshwater (riverine) water is reduced and the water turns from brackish to extremely salty.

Ice edge — The demarcation between the open sea and sea ice of any kind, whether landfast or pack.

Indicator species — A species that defines specific characteristics of its environment. Fluctuations in its population may provide an early indication of environmental change.

Indigenous — Native to an area, but can be found elsewhere as well.

Infauna — Benthic species living in the [soft] sea bottom.

Invasive species — Exotic species (i.e., alien or introduced) that rapidly establish themselves and spread through the natural communities into which they are introduced (also see *Exotic species*).

Isobath — A contour line on a bathymetric chart connecting points of equal depth in a body of water to represent sea level and depth ranges.

Keystone species — Species which are critically important for maintaining ecological processes or the diversity of their ecosystems (e.g., sea otter).

Landfast ice or Fast ice — Sea ice which forms and remains attached to the coast; it may be formed directly from sea water or by freezing of pack ice and it may extend a few metres or several hundred kilometres from the coast.

Life cycle — The entire lifespan of an organism, from fertilization to reproduction, and eventually death.

Mangrove — A tropical tree specially adapted to living in seawater, either continually or during high tides.

Marine Protected Area (MPA) — Any area of intertidal or subtidal terrain, together with its overlying waters and associated flora, fauna, historical and cultural features, which has been reserved by legislation or other effective means to protect part or all of the enclosed environment.

Moraine — Any glacially formed accumulation of unconsolidated debris which can occur in currently glaciated and formerly glaciated regions, such as those areas acted upon by a past ice age.

Nekton — Free-swimming aquatic animals, independent of the motion of waves or currents.

Neritic — The ocean environment, landward of the shelf-slope break. Also known as coastal or continental shelf waters. The neritic environment is generally less than 200 m deep (compare with *Oceanic*).

Nursery area/ground — An area in which the density of sub-adult organisms is greater than in other habitats, and in which the habitat confers advantages that result in greater survival of such organisms into the next larger size class.

Nutrient loading — The delivery of nutrients, subsequently used as food, to an area—either by runoff, river flow, atmospheric deposition, or currents.

Oceanic — Associated with marine environments seaward of the shelf break (compare with *Neritic*).

Oligotrophic — A body of water with low supply of nutrients, and subsequent low productivity as well (compare with *Eutrophic*).

Ooze (calcareous ooze) — Formed partially from the remains of small marine organisms, calcareous ooze is a fine-grained sediments of biological origin found in the deep sea.

Pacific Decadal Oscillation (PDO) — El Niño-like multi-decadal fluctuations in air and sea surface temperatures in the North Pacific affecting the North American climate. PDO events have two main periodicities: one from 15 to 25 years, the other from 50 to 70 years.

Pack ice — A term used in a broad sense to include any area of sea ice other than landfast ice. Normally applies to broken and separate blocks of ice (floes), which tend to make navigation very difficult. Occurs in various concentrations from open pack where the floes are generally not in contact with each other, to close pack ice where the ice blocks tend to be in contact.

Patch reef — Unattached to the major reef structure, a patch reef is a coral boulder or clump formed in less than 70 m depth, often in lagoons of barrier reefs or atolls.

Pelagic — Defining a habitat of or an organism that inhabits/frequents (as in seabirds) the open ocean/water column, away from the sea bottom (compare with *Benthic*).

Phocid — Hair seals of the Family *Phocidae* lacking external ears.

Photosynthesis	The process through which plants and algae convert light energy to chemical energy and store it in the chemical bonds of sugars that are the building blocks for plant biomass. Photosynthesis takes place in the chloroplasts, specifically using a light-absorbing pigment known as chlorophyll. Plants need only light energy, CO_2, H_2O, and certain nutrients to make these sugars and release oxygen as a waste product.	Seagrass	Flowering plants that colonize soft-bottom areas of the oceans from the tropics to the temperate zones.
		Seamount	A submerged mountain or volcano.
		Seashore	All ground between ordinary high-and low-water mark.
		Shoal	A stretch of shallow water, usually caused by the deposition of sediment.
Physiographic	Relating to the physical geographic features of the earth's surface: geomorphology.	Shore leads	A fracture in sea ice between pack ice and the shore. These important ice-edge systems, based on an intense bloom of microscopic plants and amphipods, are some of the richest sea areas on earth.
Plankton	Plants (phytoplankton) and animals (zooplankton) that are passively floating or weakly self-propelled.		
Polynya	Localized breaches in the ice where currents and upwelling create open water. These important ice-edge systems, based on an intense bloom of microscopic plants and amphipods, are some of the richest sea areas on earth. They provide seabirds and marine mammals with refuges in winter and feeding areas in spring and fall.	Species	1. A reproductively isolated aggregate of interbreeding organisms having common attributes and usually designated by a common name. 2. An organism belonging to such a category.
		Spur and groove structures	Alternating spurs and grooves of reef, several meters wide and up to 300 m long, appear to be formed by erosion reinforced by the prolific seaward growth of corals on grooves.
Population	A group of interbreeding organisms occupying a particular space; the number of humans or other living creatures in a designated area.	Stock	A genetically distinct population of organisms or a discreet population of fish or invertebrates targeted by a single fishery.
Primary Production	Ability of plants (in marine environments, phytoplankton, seaweeds, algae) to use the energy in sunlight to fix carbon dioxide (CO_2) into organic material, through a process known as photosynthesis. This is how plants and algae produce biomass and form the base of the food chain, as autotroph organisms (primary producers) (see *Photosynthesis*).	Subduction zone	The place where two lithosphere plates come together, one diving beneath the leading edge of the other plate. Also known as a convergent boundary.
		Sub-species	Subdivision of a species. Usually defined as a population or series of populations occupying a discrete range and differing genetically from other geographical races of the same species.
Primary Productivity	The rate at which energy accumulates in plant biomass. Its common unit of measurement is the mass of carbon fixated per unit area per year (g C/m²/yr). Since primary producers are at the foundation of their trophic pyramid, productivity can be related to the carrying capacity of an ecosystem.	Sustainable development	Development that meets the needs of the present without compromising the ability of future generations to meet their own needs.
		Terrigenous sediment	Sediment derived from land.
Pycnocline	A zone of water where the density changes rapidly with depth.	Thermocline	A narrow zone between the warm surface water and the cold water below. Temperature decline rapidly with increasing depth within the thermocline. It is also a barrier to mixing due to the densitiy differences between the upper and lower layers.
PSU (Practical Salinity Units)	One of several possible measures of salinity in water. Ocean water, for example, is generally about 35 PSU, which is equivalent to 3.5% salt content, or 35,000 ppm salt.		
Ramsar	The mission of the Ramsar Convention on Wetlands, which was signed in Ramsar, Iran, in 1971, is the conservation and wise use of all wetlands through local, regional and national actions and international cooperation, as a contribution towards achieving sustainable development throughout the world.	Trophic dynamics/ trophic structure	The movement of energy and nutrients between organisms and their environment, and through food webs in an ecosystem.
		Turbidity	A measure of the clarity of the water; high turbidity means low clarity or light penetration.
Recruitment	The amouth of fish added to the exploitable stock each year due to growth and/or migration into the fishing area. The term is also used to refer to the number of fish from a year class reaching a certain age.	Upwelling	The transport of deeper, nutrient-rich waters to the surface by wind or surface circulation patterns, which results in an increase in surface productivity. Upwelling areas are often important fishing areas.
Rhodolith beds	Rhodoliths, or "red rocks," are free-living, calcareous red algae that can form dense aggregations in shallow water (intertidal to about 200 m) from pole to pole, but are particularly abundant in subtropical waters, such as the Gulf of California. They are slow growing (less than 1 millimeter per year)—large beds are hundreds of years old. They provide habitat for numerous invertebrates and seaweeds.	Warm core rings	Eddies that detach from major currents (such as the Gulf Stream), forming a ring of self-contained water that travels, moving organisms along with it.
		Wetland	An area that is saturated by surface or ground water with vegetation adapted for life under those soil conditions, as swamps, bogs, fens, marshes, and estuaries.
Rookery	The breeding ground of certain birds or animals, such as seals.		

A blue whale in the Gulf of California. Blue whales were severely depleted during commercial whaling activities during the early 1900s in the North Pacific and along the West Coast as far south as Baja California.   *Photo:* Patricio Robles Gil

Selected References

Agardy, T. and T. Wilkinson. 2004. Conceptualizing a system of marine protected area networks for North America. In *Making ecosystem-based management work. Proceedings of the fifth international conference on science and the management of protected areas. Held at the University of Victoria, Victoria, B.C. May 11-16, 2003.* N.W.P Munro, P. Dearden, T.B. Herman, K. Beazley and S. Bondrup-Nielson, ed.Wolfville, NS: SAMPAA.

Agardy, T. and L. Wolfe. 2002. *Institutional options for integrated management of a North American marine protected areas network: a CEC report.* Montreal: CEC.

Aguayo-Lobo, A., S. Gaona, G. López-Ortega and M. Salinas-Zacarías. 1992. Mamíferos marinos, dulceacuícolas, semiacuáticos y con tendencia al agua. In *Mastofauna (mamíferos). biogeografia. naturaleza. mapa IV.8.9 Atlas nacional de México.* Instituto de Geografía. UNAM. Sistemas de Información Geográfica, S.A. México, D.F.

Ahmed, M.K. and G.M. Friedman. 1999. The impact of toxic waste dumping on the submarine environment: A case study from the New York Bight. *Northeastern Geology and Environmental Sciences* 21: 102-120.

Airamé, S., S. Gaines and C. Caldow. 2003. *Ecological linkages: marine and estuarine ecosystems of central and northern California.* Silver Spring, MD: NOAA, National Ocean Service.

Alaska Invasive Species Working Group. 2006. December audio conference. US FWS.

Alidina, H. and J. Roff. 2003. *Classifying and mapping physical habitat types (seascapes) in the Gulf of Maine and the Scotian Shelf: seascapes version to May 2003.* WWF-Canada and CLF, Gulf of Maine/Scotian Shelf MPA Planning Project.

Allen G.R. and D. R. Robertson. 1998. *Peces del Pacifico oriental tropical.* 2ª. Edicion en español. Mexico: Comision Nacional para el Conocimiento y uso de la Biodiversidad, Agrupación Sierra Madre S.C.

Alvarez-Borrego, S. 2002. Physical oceanography. In: *Island biogeography in the Sea of Cortéz,* T.J. Case, M.L. Cody, and E. Ezcurra, ed. New York: Oxford University Press.

Amezcua-Linares, F. 1996. *Peces demersales de la plataforma continental del Pacífico Central de México.* México: ICML (UNAM)-Conabio.

Angliss, R.P. and K.L. Lodge. 2002. *Alaska marine mammal stock assessments. NOAA Tech Memo. NMFS-AFSC-133.* Seattle: National Marine Mammal Laboratory, Alaska Fisheries Science Center.

Aragón Noriega, A. and L.E. Calderón Aguilera. 1997. Feasibility of intensive shrimp culture in Sinaloa, Mexico. *World Aquaculture* March 1997: 64-65.

Aragón-Noriega, E.A. and L.E. Calderon-Aguilera. 2000. Does damming the Colorado River affect the nursery area of blue shrimp *Litopenaeus stylirostris* (Decapoda: Penaeidae) in the Upper Gulf of California? *Revista de Biología Tropical/International Journal of Tropical Biology and Conservation* 48(4): 867-871.

Aragón-Noriega, E.A. and L.E. Calderon-Aguilera. 2001. Age and growth of shrimp postlarvae in the Upper Gulf of California. *Aqua Journal of Ichthyology and Aquatic Biology* 4(3): 99-104.

Arreguín-Sánchez, F. and L. E. Calderón-Aguilera. 2002. Evaluating harvesting strategies for fisheries of the ecosystem of the Central Gulf of California. In *The Use of Ecosystem Models to Investigate Multispecies Management Strategies for Capture Fisheries, FAO - Fisheries Center Research Report* 10(2): 135-141. T. Pitcher and K. Cochrane, eds.

Arriaga-Cabrera, L., E. Vázquez Domínguez, J. González-Cano, R. Jiménez-Rosemberg, E. M Muñoz-López, V. Aguilar-Sierra, coords. 1998. *Regiones prioritarias marinas de México.* México: Comisión Nacional para el Conocimiento y uso de la Biodiversidad.

Bakun, A. 1993. The California Current, Benguela Current, and Southwestern Atlantic shelf ecosystems: A comparative approach to identifying factors regulating biomass yields. In *Large marine ecosystems, stress, mitigation and sustainability,* K. Sherman, L. M. Alexander, and B. D. Gold ed., 199-221. Washington, D.C.: American Association for the Advancement of Science.

Beck, M., K.L. Heck, K.W. Able, D.L. Childers, D.B. Eggleston, B.M. Gillanders, B. Halpern, C.G. Hays, K. Hoshino, T.J. Minello, R.J. Orth, P.F. Sheridan, and M.P. Weinstein. 2001. The identification, conservation, and management of estuarine and marine nurseries for fish and invertebrates. *BioScience* 51(8): 6-33-641.

Bernardi, G., L. Findley and A. Rocha-Olivares. 2003. Vicariance and dispersal across Baja California in disjunct marine fish populations. *Evolution* 57(7): 1599-1609.

Bezaury-Creel J., R. Macias Ordoñez, G. García Beltrán, G. Castillo Arenas, N. Pardo Caicedo, R. Ibarra Navarro, A. Loreto Viruel. 1997. Implementation of the International Coral Reef Initiative (ICRI) in Mexico. In *The International Coral Reef Initiative: the status of coral reefs in Mexico and the United States Gulf of Mexico.* Compact Disk, Amigos de Sian Ka'an A.C., CINVESTAV, NOAA, CEC, The Nature Conservancy.

Bezaury Creel J., A. Mosso, D.Gutierrez. 1996. *Estrategia para la conservación de areas costeras y marinas para México. Borrador para discusión.* Montreal, Canada: Documento Interno Amigos de Sian Ka'an A.C. UICN – CNPPA.

Bianchi, T.S., J.R. Pennock and R. R. Twilley, eds. 1999. *Biogeochemistry of Gulf of Mexico estuaries.* New York: John Wiley & Sons Inc.

Birkett, S.H. and D.J. Rapport. 1999. A Stress-Response Assessment of the Northwestern Gulf of Mexico Ecosystem. In: *The Gulf of Mexico Large Marine Ecosystem: assessment, sustainability, and management* H. Kumpf, H.K. Steidinger and K. Sherman, eds. Malden Mass: Blackwell Science.

Bleakley, C. and V. Alexander. 1995. Marine Region 2 Arctic. In *A global representative system of marine protected areas. Vol. 1. Antarctic, Arctic, Mediterranean, Northwest Atlantic, Northeast Atlantic and Baltic,* 61–76. Washington, D.C.: Great Barrier Reef Marine Park Authority, World Bank, World Conservation Union (IUCN).

Botello, A.V., J.L. Rojas Galaviz, J. Benitez y D. Zárate Lomelí, ed. 1996. *Golfo de México Contaminación e Impacto Ambiental: Diagnóstico y Tendencias. UAC-EPOMEX Serie Científica 5.* Campeche: UAC-EPOMEX.

Bottom, D. L., K.K. Jones, J.D. Rodgers and R.F. Brown. 1993. Research and management in the Northern California Current ecosystem. In *Large marine ecosystems, stress, mitigation and sustainability,* K. Sherman, L.M. Alexander, and B.D. Gold, eds., 259-271. Washington, D.C.: American Association for the Advancement of Science.

Briggs, J. C. 1974. *Marine zoogeography.* New York: NYT, McGraw-Hill Book Co.

Briggs J.C. 1995. *Global biogeography.* Elsevier Science B.V.

Brodeur R.D. and D.M. Ware. 1995. Interdecadal variability in distribution and catch rates of epipelagic nekton in the northeast Pacific Ocean. In *Climate change and northern fish populations,* R. Beamish, ed., 329-356. *Can. Spec. Pub. Fish. Aquat. Sci. 121.*

Brown, S. K., R. Mahon, K.C.T. Zwanenburg, K.R. Buja, L.W. Claflin, R.N. O'Boyle, B. Atkinson, M. Sinclair, G. Howell and M.E. Monaco. 1996. *East coast of North America groundfish: initial explorations of biogeography and species assemblages.* Silver Spring, MD and Dartmouth, NS: National Oceanic and Atmospheric Administration and Department of Fisheries and Oceans.

Brusca, R.C., L.T. Findley, P.A. Hastings, M.E. Hendrickx, J. Torre Cosio and A.M. van der Heiden. 2005. Macrofaunal diversity in the Gulf of California. In: *Biodiversity, ecosystems and conservation in northern Mexico*, J.-L. Cartron, G. Ceballos, and R.S. Felger, eds., 179-203. New York: Oxford University Press.

Brusca R.C., E. Kimrey and W. Moore. 2004. *A seashore guide to the northern Gulf of California*. Arizona. Tucson, Arizona: Sonora Desert Museum.

CAFF (Conservation of Arctic Flora and Fauna). 2004. *Arctic flora and fauna: recommendations for conservation*. Helsinki, Finland: Edita Plc.

Calderon Aguilera, L.E. 1992. Analysis of the benthic infauna from Bahia de San Quintín, Baja California, with emphasis on its use in impact assessment. *Ciencias Marinas* 18(4): 2746.

Calderon-Aguilera, L.E. 1998. Past and present features of Strait of Georgia macrobenthos. In: *Back to the future: reconstructing the Strait of Georgia ecosystem, Fisheries Centre Research Reports* 6(5): 38-41, D. Pauly, T. Pitcher and D. Preikshot, ed. Vancouver: The Fisheries Centre, University of British Columbia.

Calderon-Aguilera L.E., E. A. Aragón-Noriega, H. A. Licón, G. Castillo-Moreno and A. Maciel-Gómez. 2002. Abundance and composition of penaeid postlarvae in the Upper Gulf of California. In *Contribution to the study of East Pacific crustaceans*, M.E. Hendrickx, ed., 281-292. México: Inst. Cienc. Mar. Limnol., UNAM.

Calderon-Aguilera L.E., S.G. Marinone and E.A. Aragón-Noriega. 2003. Influence of oceanographic processes on the early life stages of the blue shrimp (*Litopenaeus stylirostris*) in the upper Gulf of California. *Journal of Marine Systems* 39: 117-128.

California Dept. of Fish and Game. 1990. Review of some California fisheries for 1989. *CalCOFI Rep.* 31: 9-21.

Cameron, C. and S. Mitchell. 2000. *St. Georges Bay ecosystem project. Research report II: Diets and feeding.* Nova Scotia: Interdisciplinary Studies in Aquatic Resources, St. Francis Xavier University, Antigonish, Nova Scotia for Fisheries and Oceans Canada, Can. Manuscr. Rep. Fish Aquat. Sci. 2512.

Carranza-Edwards, A. 1986. Estudio sedimentológico regional de playas del estado de Chiapas, México. *An. Inst. Cienc. del Mar y Limnol., Univ. Nal. Autón. México* 13(1): 331-344.

Carranza-Edwards, A. 2001. Grain size and sorting in modern beach sands. *Journal of Coastal Research* 17 (1): 38-52.

Carranza-Edwards, A. and J.E. Aguayo-Camargo. 1992. *Sedimentología marina (Carta de Geología Marina, Esc. 1:12,000,000). Vol. II – Naturaleza - Hoja IV.9.5.B del Atlas Nacional de México.* Instituto de Geografía, UNAM.

Carranza-Edwards, A., G. Bocanegra-García, L. Rosales-Hoz and L. De Pablo Galán. 1996. Beach sands from Baja California Peninsula, Mexico. *Sedimentary Geology* 119: 263-274.

Carranza-Edwards, A., E. Centeno-García, L. Rosales-Hoz and R. Lozano Santa Cruz. 2001. Provenance of beach gray sands from western México. *Journal of South American Earth Sciences* 14: 291-305.

Carranza-Edwards, A., M. Gutérrez-Estrada and R. Rodríguez-Torres. 1975. Unidades morfotectónicas continentales de las costas mexicanas. *A. Centro Cienc. del Mar y Limnol. UNAM* 2(1): 81-88.

Carranza-Edwards, A., A.Z. Márquez-García and E.A. Morales de la Garza. 1986. Estudio de sedimentos de la plataforma continental del estado de Guerrero y su importancia dentro de los recursos minerales del mar. *An. Inst. Cienc. del Mar y Limnol., Univ. Nal. Autón. México*, 13(3): 241-262.

Carranza-Edwards, A., A.Z. Márquez-Garcia and E.A. Morales de la Garza. 1987. Distribución y características físicas externas de nódulos polimetálicos en el sector central del Pacífico Mexicano. *Bol. Mineral.*3(1): 78-94.

Carranza-Edwards A., E. Morales de la Garza and L. Rosales Hoz. 1998. Tectonics, sedimentology and geochemistry. Chap. 1. In: *El Golfo de Tehuantepec: el ecosistema y sus recursos.* M. Tapia-García ed., 1-12. Iztapalapa, México: Universidad Autónoma Metropolitana-Iztapalapa.

Carranza-Edwards, A. and L. Rosales-Hoz. 1994. Importancia de los recursos minerales marinos de la región. In: *La Isla Socorro, Reserva de la Biosfera, Archipiélago de las Revillagigedo, México,* Ortega-Rubio, A. y A. Castellanos-Vera, eds, 33-42. La Paz: Centro de Investigaciones Biológicas de La Paz, Publicación No. 8 del Centro de Investigaciones Biológicas del Noroeste, S. C., en colaboración con el Fondo Mundial de la Vida Silvestre (WWF).

Carranza-Edwards, A. and L. Rosales-Hoz, 1995. Grain-size trends and provenance of southwestern Gulf of Mexico beach sands. *Can. J. Earth Sci.* 32: 2009-2014.

Carranza-Edwards, A., L. Rosales-Hoz, J.E. Aguayo-Camargo, Y. Hornelas-Orozco and R. Lozano-Santa Cruz. 1990. Geochemical study of hydrothermal core sediments and rocks from the Guaymas Basin, Gulf of California. *Applied Geochemistry* 5: 77-82.

Carranza-Edwards, A., L. Rosales-Hoz, A. Aguirre-Gómez and A. Galán-Alcalá. 1988. Estudio de metales en sedimentos litorales de Sonora, México. *An. Inst. Cienc. del Mar y Limnol., Univ. Nal. Autón. México* 15(2): 225-234.

Carranza-Edwards, A., L. Rosales-Hoz and R. Lozano-Santa Cruz. 1988. Estudio sedimentológico de playas del Estado de Oaxaca, Mexico. *An. Inst. Cienc. del Mar y Limnol., Univ. Nal. Autón. México* 15(2): 23-38.

Carranza-Edwards, A., L. Rosales-Hoz and M.A. Monreal-Gómez. 1993. Suspended sediments in the southeastern Gulf of Mexico. *Mar. Geol.* 112: 257-269.

Carranza-Edwards, A., L. Rosales-Hoz, E. Ruíz-Ramírez and S. Santiago-Pérez. 1989. Investigations of phosphorite deposits in the Gulf of Tehuantepec. *Marine Mining* 8: 317-323.

Carranza-Edwards, A., L. Rosales-Hoz and S. Santiago-Pérez. 1994. Provenance memories and maturity of holocene sands in Northwest Mexico. *Can. J. Earth Sci.* 31: 1550-1556.

Carranza-Edwards, A., L. Rosales-Hoz and S. Santiago-Pérez. 1996. A reconnaissance study of carbonates in Mexican beach sands. *Sedimentary Geology* 101: 261-268.

Case, T. and M. Cody, ed. 1983. Appendix. Fig. 3.3 Near-surface temperatures. *Island Biogeography of the Sea of Cortéz.* Berkeley, California: University of California Press.

Castañeda, O. and F. Contreras. 1995. *Ecosistemas costeros Mexicanos.* México, D.F.: Conabio - UAM Iztapalapa. Publicaciones Electrónicas de México, S.A. de C.V.

Catto, N.R., M.R. Anderson, D.A Scruton and U.P. Williams. 1997. *Coastal classification of the Placentia Bay shore.* Canadian Technical Report of Fisheries and Aquatic Sciences No. 2186, St. John's, Nfld: Department of Fisheries and Oceans.

CEC. 1997. *Ecological regions of North America: toward a common perspective.* Montreal, Quebec: Commission for Environmental Cooperation.

Christopher, F.G. Jeffrey, R. Anlauf, J. Beets, S. Caseau, W. Coles, A.M. Friedlander, S. Herzlieb, Z. Hillis-Starr, M. Kendall, V. Mayor, J. Miller, R. Nemeth, C. Rogers and W. Toller 2005. The state of coral reef ecosystems of the US Virgin Islands In *The state of coral reef ecosystems of the United States and Pacific Freely Associated States: 2005. NOAA technical memorandum NOS NCCOS 11.* J. Waddell, ed., 45-90. Silver Spring, MD: NOAA/NCCOS Center for Coastal Monitoring and Assessment's Biogeography Team.

Clark, A.M. and D. Gulko. 1999. *Hawaii's state of the reefs report, 1998.* Honolulu: Department of Land and Natural Resources.

Clark, J.R. 1982. *Assessing the National Estuary Sanctuary Program, action summary. Report to the Estuary Sanctuary Program Manger, Office of Coastal Zone Management.* Purchase Order No. NA81AAA03317, March 9, 1982.

Conanp. 2004. *Programa de conservación y manejo Reserva de la Biosfera Archipiélago de Revillagigedo.* México: Secretaría de Medio Ambiente Recursos Naturales y Pesca.

Contreras, F. 1985. *Las lagunas costeras Mexicanas.* México: Centro de Ecodesarrollo. Secretaría de Pesca.

Contreras - Espinosa, F. 1993. *Ecosistemas costeros Mexicanos.* Iztapalapa, México: Conabio - UAM Iztapalapa.

Cortés, R., ed. 2002. *Latin American coral reefs.* Amsterdam: Elsevier Science Ltd.

COSEWIC 2003. *COSEWIC assessment and update status report on the Atlantic cod* Gadus morhua *in Canada.* Ottawa: Committee on the Status of Endangered Wildlife in Canada.

Crawford, W.R. 1997. Physical oceanography of the waters around the Queen Charlotte Islands. Pp. In *The ecology, status, and conservation of marine and shoreline birds of the Queen Charlotte Islands. Occasional Paper No. 93,* K. Vermeer and K.H. Morgan, ed., 8–17. Delta, B.C.: Canadian Wildlife Service, Environment Canada.

Croom, M., R. Wolotira and W. Henwood. 1995. Marine Region 15 Northeast Pacific. In *A global representative system of marine protected areas. Vol. IV. South Pacific, Northeast Pacific, Northwest Pacific, Southeast Pacific and Australia/New Zealand,* 55–106. Washington, D.C.: Great Barrier Reef Marine Park Authority, World Bank, World Conservation Union (IUCN).

Dailey, M.D., D.J. Reish and J.W. Anderson, ed. 1994. *Ecology of the Southern California Bight: a synthesis and interpretation.* Berkeley, CA: University of California Press.

Day, J.C. and J.C. Roff. 2000. *Planning for representative marine protected areas: a framework for Canada's oceans.* Toronto: World Wildlife Fund Canada.

De la Lanza-Espino, G. 1991. *Oceanografía de mares Mexicanos.* México: A.G.T. Editor S.A.

Dickens, D., I. Bjerkelund, P. Vonk, S. Potter, K. Finley, R. Stephen, C. Holdsworth, D. Reimer, A. Godon, I. Buist and A. Sekerak. 1990. *Lancaster Sound Region—A coastal atlas for environmental protection.* Yellowknife, N.W.T.: Environment Canada.

Drazen, J.C., S.K. Goffredi, B. Schlining and D.S. Stakes. 2003. Aggregations of egg-brooding deep-sea fish and cephalopods on the Gorda Escarpment: a reproductive hot spot. *Biological Bulletin* 205:1-7.

Dreckman K. and A. Gamboa-Contreras. 1998. Ficoflora marina bentónica actualizada del Golfo de Tehuantepec y algunos registros para Guatemala. Chap. 7: In *El Golfo de Tehuantepec: el ecosistema y sus recursos.* M. Tapia-García ed., 75-91. Iztapalapa, México: Universidad Autónoma Metropolitana-Iztapalapa.

Ecoregions Working Group. 1989. *Ecoclimatic regions of Canada, first approximation. Ecological Land Classification Series No. 23,* Ottawa: Sustainable Development Branch, Environment Canada.

Ecosystem Stratification Working Group. 1995. *A national ecological framework for Canada.* Ottawa. Agriculture and Agri-Food Canada, Environment Canada.

Emmett, B., L. Burger and Y. Carolsfeld. 1995. *An inventory and mapping of subtidal biophysical features of the Goose Islands, Hakai Recreation Area, British Columbia. Occasional Paper No. 3.* Victoria: B.C. Ministry of Environment, Lands and Parks.

EPA. 2005. *National coastal condition report II.* Washington: US Environmental Protection Agency.

Estes, J. A. and D.O. Duggins. 1995. Sea otters and kelp forests in Alaska: generality and variation in a community ecological paradigm. *Ecological Monographs* 65(1): 75-100.

Estrada-Ramírez, A. and L.E. Calderon-Aguilera. 2001. Range extension for *Sicyonia penicillata* on the western coast of Baja California, Mexico. *Crustaceana* 74(3): 317-320.

Etnoyer, P., D. Canny, B. Mate, and L. Morgan (2004). Persistent pelagic habitats in the Baja California to Bering Sea (B2B) Ecoregion. *Oceanography* 17(1): 90-101.

Fernández-Eguiarte, A., A. Gallegos-García and J. Zavala-Hidalgo. 1989. Batimetría (carta de hipsometría y batimetría, esc. 1:4,000,000), *Vol. I – Mapas generales - hoja I.1.1 del atlas nacional de México.* México: Instituto de Geografía, UNAM.

Fernández-Eguiarte, A., A. Gallegos-García and J. Zavala-Hidalgo. 1992. Aspectos regionales de los mares mexicanos. In. *Oceanografía física 2. oceanografía. naturaleza. mapa IV.9.2 atlas nacional de México.* Instituto de Geografía. UNAM. Sistemas de Información Geográfica, S.A., ed. México: Instituto de Geografía. UNAM. Sistemas de Información Geográfica, S.A.

Fernández-Eguiarte, A., A. Gallegos-García and J. Zavala-Hidalgo. 1992. Masas de agua y mareas de los mares mexicanos. In. *Oceanografía física 1. oceanografía. naturaleza. mapa IV.9.1 atlas nacional de México.* Instituto de Geografía. UNAM. Sistemas de Información Geográfica, S.A., ed. México: Instituto de Geografía. UNAM. Sistemas de Información Geográfica, S.A.

Flores, D., P. Sánchez-Gil, J.C. Seijo and F. Arreguín, ed. 1997. *Análisis y Diagnóstico de los recursos pesqueros críticos del Golfo de México. UAC-EPOMEX Serie Científica 7.* Campeche: UAC-EPOMEX. OF FI

Ford, J.R. and M.L. Bonnell. 1996. *Developing a methodology for defining marine bioregions: The Pacific Coast of the Continental US.* Report to the World Wildlife Fund (WWF). Portland, OR.: Ecological Consulting Inc., WWF.

Foster, M. S. and D. R. Schiel. 1985. *The ecology of giant kelp forests in California: a community profile.* US Fish and Wildlife Service Biological Report 85(7.2).

Freeland, H. 1992. The physical oceanography of the west coast of Vancouver Island. In *The ecology, status and conservation of marine and shoreline birds on the west coast of Vancouver Island. Occasional Paper No. 75.* K. Vermeer, R.W. Butler, K.H. Morgan, ed., 10–14. Delta, B.C.: Canadian Wildlife Service, Environment Canada.

Friedlander, A.M. and E.E. DeMartini. 2002. Contrasts in density, size, and biomass of reef fishes between the northwestern and main Hawaiian Islands: the effects of fishing down apex predators. *Mar. Ecol. Prog. Ser.* 11:1-14.

Friedlander, A.M., G. Aeby, E. Brown, A. Clark, S. Coles, S. Dollar, C. Hunter, P. Jokiel, J. Smith, B. Walsh, I. Williamsa and W. Wiltse. 2005. The state of coral reef ecosystems of the Main Hawaiian Islands. In *The state of coral reef ecosystems of the United States and Pacific Freely Associated States: 2005. NOAA Technical Memorandum NOS NCCOS 11.* J. Waddell ed., 222-269. Silver Spring, MD: NOAA/NCCOS Center for Coastal Monitoring and Assessment's Biogeography Team.

Gallegos-García A., J.M. Barberán-Falcón, A. Fernández-Eguiarte. 1988, *Condiciones oceánicas alrededor de isla Socorro, archipiélago de Revillagigedo, en julio de 1981. Revista geofísica no. 28, Enero – Junio 1988.* Instituto Panamericano de Geografía e Historia.

Gamboa-Contreras A. and M. Tapia-García. 1998. Invertebrados bentónicos de la plataforma continental interna. Chap. 9. In *El Golfo de Tehuantepec: el ecosistema y sus recursos,* M. Tapia-García ed., 103-128. Iztapalapa, México: Universidad Autónoma Metropolitana-Iztapalapa.

Goodyear, C. P. 1996. *Status of the Red Drum Stocks of the Gulf of Mexico.* NMFS, Miami Laboratory Contribution No. MIA-95/96-47.

Gutiérrez, C. and J. Bezaury. 2001. *Mapeo de ecosistemas marinos y estuarinos de Norte América (México), Documento de trabajo.* Comisión de Cooperación Ambiental, Fondo Mundial para la Naturaleza, WWF-México.

Gutiérrez, D., C. García - Saez, M. Lara and C. Padilla. 1993. Comparación de arrecifes coralinos: Veracruz y Quintana Roo. In *Biodiversidad marina y costera de México.* S.I. Salazar - Vallejo y N.E. González, ed., 787 - 806. México: Conabio y CIQRO.

Hanson, A.J., T. Agardy and R. Perez Gil Salcido. 2000. *Securing the continent's biological wealth: toward effective biodiversity conservation in North America. A working draft.* Montreal: CEC.

Harding, L.E. and H. Hirvonen. 1996. A marine ecosystem classification system for Canada. In *Proceedings, marine ecosystem monitoring network workshop, Nanaimo, B.C., March 28–30, 1995.* D.E. Hay, R.D. Waters, and T.A. Boxwell, ed., 8–19.Canadian Technical Report of Fisheries and Aquatic Sciences 2108. Nanaimo, B.C.: Department of Fisheries and Oceans.

Harper, J.R., J. Christian, W.E. Cross, R. Frith, G. Searing and D. Thompson. 1993. *Final report. A classification of the marine regions of Canada.* Sidney, B.C.: Coastal and Ocean Resources Inc.

Harper, J.R., G.A. Robilliard and J. Lathrop. 1983. *Marine regions of Canada: framework for Canada's system of national marine parks. Final report to Parks Canada.* Victoria, B.C.: Woodward-Clyde Consultants.

Hatch, S. A. and G. A. Sanger. 1992. Puffins as samplers of juvenile pollock and other forage fish in the Gulf of Alaska. *Mar. Ecol. Progress Series* 80: 1-14.

Hayden B.P., G. Carleton Ray, R. Dolan. 1984. Classification of coastal and marine environments. In *Environmental Conservation* 11(3): 109-207.

Hendrickx, M.E., R.C. Brusca and L.T. Findley. 2005. Listado y Distribución de la Macrofauna del Golfo de California, México, Parte 1, Invertebrados (A Distributional Checklist of the Macrofauna of the Gulf of California, Mexico, Part 1, Invertebrates). 429 pp. Tucson: Arizona-Sonora Desert Museum (and Guaymas, Sonora: Conservación Internacional–Región Golfo de California)

Howes, D.E. 2000. *Science and information initiatives for marine protected areas in British Columbia.* Victoria, B.C.: Land Use Coordination Office.

Howes, D.E. 2001. BC biophysical shore-zone mapping system – a systematic approach to characterize coastal habitats in the Pacific Northwest. Victoria, B.C.: Land Use Coordination Office.

Howes, D.E. and P. Wainwright. 1999. *Coastal resource and oil spill response atlas for west coast Vancouver Island.* CD ROM Atlas. Victoria, B.C.: Province of British Columbia.

Howes, D.E., M.A. Zacharias and J.R. Harper. 1997. *British Columbia marine ecological classification: marine ecosections and ecounits.* Victoria, B.C.: Province of British Columbia.

Hunt, G.L., Jr., P. Stabeno, G. Walters, E. Sinclair, R.D. Brodeur, J.M. Napp and N.A. Bond. 2002. *Climate change and control of the southeastern Bering Sea pelagic ecosystem.* Deep-Sea Research Part II: Topical Studies in Oceanography: Ecology of the Southeastern Bering Sea.

IUCN/UNEP/WWF. 1991. *Caring for the earth. A strategy for sustainable living.* Gland, Switzerland: IUCN/UNEP/WWF.

Jacobs, D.K., T.A. Haney and K.D. Louie. 2004. Genes, diversity, and geologic process on the Pacific Coast. *Annual Review of Earth and Planetary Sciences* 32: 601-652.

Jamieson, G.S., C.O. Levings. 2001 Marine protected areas in Canada: implications for both conservation and fisheries management. *Canadian J. Fisheries and Aquatic Sciences* 58: 138-156.

Jaramillo Legorreta, A.M., L. Rojas Bracho and T. Gerrodette. 1999. A new abundance estimate for vaquitas: first step for recovery. *Marine Mammal Science* 15(4): 957-973.

Johns, G. M., V. R. Leeworthy, F. W. Bell and M. A. Bonn. 2001. *Socio-economic study of reef resources in southeast Florida and the Florida Keys.* Fort Lauderdale, Florida: Broward County Department of Planning and Environmental Protection.

Kasper-Zubillaga, J., A. Carranza-Edwards and L.Rosales-Hoz. 1999. Petrography and geochemistry of holocene sands in the western Gulf of Mexico: implications for provenance and tectonic setting. *Journal of Sedimentary Research* 69(5): 1003-1010.

Kelleher, G., C. Bleakley, and S. Wells, ed. 1995. *A global representative system of marine protected areas.* 4 vols.Washington, D.C.: Great Barrier Reef Marine Park Authority, World Bank, World Conservation Union (IUCN).

Kumpf, H., K. Steidinger and K. Sherman, ed. 1999. *The Gulf of Mexico large marine ecosystem: assessment, sustainability and management.* Malden, Mass.: Blackwell Science Inc.

Lancin, M. and A. Carranza. 1976. Estudio geomorfológico de la Bahía y de la Playa de Santiago en Manzanillo, Colima. *Inst. Geol., Univ. Nal. Autón. México, Rev.* 2: 43-65.

Lara M., D. Gutiérrez, J. Bezaury, C. Padilla and R.M. Loreto. 2001. Management and monitoring proposal for the Mexican Meso-American Barrier Reef System (MBRS). In *North American and European perspectives on ocean and coastal policy: building partnerships and expanding the technological frontier, proceedings, volume 1, International conference on coastal and ocean space utilization, COSU 2000, November 1-4, 2000,* 17-24. Cancún, México: COSU.

Lara-Domínguez, A.L., A. Yáñez-Arancibia and J. W. Day. 2002. Sustainable management of mangroves in Southern Mexico and Central América, In *Managing forest ecosystems for sustainable livelihoods.* The Hague, Netherlands: The Global Biodiversity Forum.

Lara-Lara R., E. Robles-Janero, M.C. Bazán-Guzmán and E Millán Núñez. 1998. Productividad de fitoplancton. Chap. 5: In *El Golfo de Tehuantepec: el ecosistema y sus recurso,* M. Tapia-García, ed., 51-74. Iztapalapa, México: Universidad Autónoma Metropolitana-Iztapalapa.

Leet, W.S., C. M. Dewees, R. Klingbeil and E.J. Larson, ed. 2001. *California's living marine resources: a status report.* California Department of Fish and Game.

Lien, J., M. Dunn, E. Wiken and M. Padilla. 2001. *The status of Canada's oceanic and coastal habitats.* Wildlife Habitats. Scientific/Technical Report. Ottawa, Ontario: Wildlife Habitat Canada.

Lluch-Cota, S.E., S. Alvarez-Borrego, E.M. Santamaria-Del-Angel, F.E. Muller-Karger, and S. Hernandez-Vazquez. 1997. The Gulf of Tehuantepec and adjacent areas: spatial and temporal variation of satellite-derived photosynthetic pigments. *Ciencias Marinas* 23: 329-340.

Longhurst, A. R. 1998. *Ecological geography of the sea.* San Diego, California: Academic Press.

Longhurst, A. R. and D. Pauly. 1987. *Ecology of tropical oceans.* New York: Academia Press Inc..

Lowry, L.F., V.N. Burkanov and K.J. Frost. 1996. Importance of walleye pollock (*Theragra chalcogramma*) in the diet of Phocid seals in the Bering Sea and northwestern Pacific Ocean. In, *Ecology of juvenile walleye pollock,* Theragra chalcogramma, *US Department of Commerce, NOAA technical report*. R.D. Brodeur, ed., 141-151. Silver Spring, MD: NMFS.

Lozano-Santa Cruz, R., P. Altuzar-Coello, A. Carranza-Edwards and L.Rosales-Hoz. 1989. Distribución de minerales en la fracción arcillosa de sedimentos del Pacífico Central Mexicano. *An. Inst. Cienc. del Mar y Limnol., Univ. Nal. Autón. México* 16(2): 321-330.

Lugo-Hubp J. 1985. Morfoestructuras del Fondo Marino Mexicano. *Boletín del Instituto de Geografía de la Universidad Nacional Autonoma de México* 15: 9-39.

MacCall, A.D. 1986. Changes in the biomass of the California Current ecosystem. In *Variability and management of large marine ecosystems,* K. Sherman and L.M. Alexander, ed., 33-54. Boulder: Westview-AAAS Selected Symposium 99.

Macklin, S.A. 1999. *Southeast Bering Sea carrying capacity program: final report of phase I research, August 1996 – September 1998.* Silver Spring, MD: NOAA Coastal Ocean Program.

Maluf L.Y. 1983. The physical oceanography. In *Island biogeography of the Sea of Cortéz*, T. Case and M. Cody, ed., 26 – 45. Berkeley, CA: University of California Press.

Maragos, J.E. and D. Gulko, ed. 2002. *Coral reef ecosystems of the Northwestern Hawaiian Islands: interim results emphasizing the 2000 surveys.* Honolulu: US Fish and Wildlife Service and the Hawaii Department of Land and Natural Resources.

Marine Environmental Quality Advisory Group. 1994. *Marine ecological classification system for Canada.* Hull, QC: Environment Canada.

Márquez-Garcia, A.Z., A. Carranza-Edwards and E.A. Morales de la Garza. 1988. Características sedimentológicas de playas de la Isla Clarion, Colima, México. *An Inst. Cienc. del Mar y Limnol., Univ. Nal. Autón. México* 15(2): 39-48.

McGowan, J.A., D.B. Chelton and A. Conversi. 1999. Plankton patterns, climate, and change in the California Current. In *Large marine ecosystems of the Pacific Rim—assessment, sustainability, and management*, K. Sherman and Q. Tang ed., 63-105. Malden, Mass.: Blackwell Science.

Mercier, F. and C. Mondor. 1995. *Sea to sea to sea: Canada's national marine conservation areas system plan*, Ottawa: Parks Canada, Department of Canadian Heritage.

Merino-Ibarra, M. 1992. *Afloramiento en la plataforma de Yucatán: estructura y fertilización. Tesis de doctorado.* Quintana Roo: Instituto de Ciencias del Mar y Limnología, UNAM.

Merrick, R.L., T.R. Loughlin and D.G. Calkins. 1987. Decline in abundance of the northern sea lion, *Eumetopias jubatus*, in Alaska, 1956-86. *Fisheries Bulletin US* 85(2): 351-365.

Michel, C., R. G. Ingram, L. R. Harris. 2006. Variability in oceanographic and ecological processes in the Canadian Arctic Archipelago. Progress in Oceanography 71 (2006), pp. 379-401.

Miller, P. 2000. Tracking dioxins to the Arctic: a CEC study tracks dioxins from Canada, Mexico and the United States to the Arctic. In *Trio fall edition*, CEC, ed. Montreal, Canada: CEC.

Moch, A. and G.M. Friedman. 1999. The impact of organic-rich waste released into New York Bight sediment. *Northeastern Geology and Environmental Sciences* 21: 49-101.

Molina-Cruz and M. Martínez López. 1994. Oceanography of the Gulf of Tehuantepec, Mexico, indicated by Radiolaria remains. *Palaeogography, Palaeoclimatology, Palaeoecology* 110: 179-195.

Mondor, C., F. Mercier, M. Croom and R. Wolotira. 1995. Marine region 4 Northwest Atlantic. In *A global representative system of marine protected areas. V.1. Antarctic, Arctic, Mediterranean, Northwest Atlantic, Northeast Atlantic and Baltic,* G. Kelleher, G. C. Bleakley and S. Wells, ed. Washington, D.C.: Great Barrier Reef Marine Park Authority, International Bank for Reconstruction and Development/World Bank, World Conservation Union.

Morgan, L. P. Etnoyer, T. Wilkinson, H. Herrmann, F. Tsao and S. Maxwell. 2004. Identifying priority conservation areas from Baja California to the Bering Sea. In *Making ecosystem-based management work, Proceedings of the fifth international conference on science and the management of protected areas. Held at the University of Victoria, Victoria, B.C. May 11-16, 2003.*N.W.P Munro, P. Dearden, T.B. Herman, K. Beazley and S. Bondrup-Nielson, ed. Wolfville, NS: SAMPAA.

Morgan, L S. Maxwell, F. Tsao, T.A.C. Wilkinson, P. Etnoyer. 2005. *Marine priority conservation areas: Baja California to the Bering Sea.* Montreal: Commission for Environmental Cooperation – Marine Conservation Biology Institute.

Monreal Gomez, M.A. and D.A. Salas de Leon. 1998. Dynamic and thermohaline structure. Chap. 2: In *El Golfo de Tehuantepec: el ecosistema y sus recursos*, M. Tapia-García, ed., 13-26. Iztapalapa, México: Universidad Autónoma Metropolitana-Iztapalapa.

Moreno-Casasola, P. 1999. Dune vegetation and its biodiversity along the Gulf of Mexico, a large marine ecosystem. Chap. 34. In *The Gulf of Mexico large marine ecosystem: assessment, sustainability and management.* H Kumpf, K. Steidinger and K. Sherman ed., 593-612. Malden Massachusetts: Blackwell Science Inc.

Moreno-Casasola, P. and I. Espejel. 1986. Classification and ordination of coastal sand dunes vegetation along the Gulf of Mexico and Caribbean Sea, Mexico. *Vegetation* 66: 147-182.

Nelson, J.S., E.J. Crossman, H. Espinosa-Perez, L.T. Findely, C.R. Gilbert, R.N. Lea, and J.D. Williams. 2004. Common and scientific names of fishes from the United States, Canada, and Mexico, Sixth Edition. American Fisheries Society. Special publication 29. 368 pp.

NMFS. 2004. *Annual report to Congress on the status of US fisheries – 2003.* Silver Spring, MD US Dept.Commerce, NOAA, Natl. Mar. Fish. Serv.

NMFS and US FWS. 1998. *Recovery plan for the US Pacific populations of the East Pacific green turtle (Chelonia mydas).* Silver Spring, MD: NMFS.

NOAA. 1988. *Bering, Chukchi and Beaufort Seas: coastal and ocean zones. Strategic assessment: data atlas.* Silver Spring, MD: NOAA.

NOAA. 1998. *Biogeorgraphic Regions of the NERRS.* Silver Spring, MD: NOAA.

Nolasco-Montero, E. and A. Carranza-Edwards. 1988. Estudio sedimentológico regional de playas de Yucatán y Quintana Roo, México. *An. Inst. Cienc. del Mar y Limnol., Univ. Nal. Autón. México* 15(2): 49-66.

Norse, E.A., ed. 1993. *Global marine biological diversity: a strategy for building conservation into decision making.* Washington, DC: Island Press.

NRC. 2003. *Ocean noise and marine mammals.* Washington, DC: National Academies Press.

National Wetlands Working Group. 1987. *The Canadian wetland classification system, provisional edition. Ecological Land Classification Series. No. 21.* Ottawa: Canadian Wildlife Service, Environment Canada.

Oceans Ltd. 1994. *Reproductive success of groundfish stocks in the Canadian Northwest Atlantic.* Ottawa: Atlantic Stock Assessment Secretariat. Ottawa: Department of Fisheries and Oceans.

Ortiz-Pérez, M.A. and L.M. Espinosa-Rodríguez. 1992. *Geomorfología 2. relieve. tipos de costas. hoja IV.3.4.D. (Esc. 1:8,000.000) del atlas nacional de México.* México: Instituto de Geografía, UNAM.

Ortiz Pérez, M.A., C. Valverde and N.P. Psuty. 1996. The impacts of sea-level rise and economic development on the low-lands of the Mexican gulf coast. *Golfo de México contaminación e impacto ambiental: diagnóstico y tendencias, UAC-EPOMEX Serie Científica 5:* 459-470.

Owens, E.H. 1994. *Canadian coastal environments, shoreline processes, and oil spill cleanup. EPS 3/SP/5,* Ottawa: Environment Canada.

Pearce, J.B. The New York bight. *Marine Pollution Bulletin* 41: 44-55.

Pérez-Peña, M. and E. Ríos-Jara. 1998. Gastropod mollusks from the continental shelf off Jalisco and Colima, Mexico. *Ciencias Marinas* 24(4): 425-442.

Pitt, J. 2001. Can we restore The Colorado River Delta? *Journal of Arid Environments* 49: 211-220.

Ray, G.C., B.P. Hayden and R. Dolan. 1982. *Development of a biophysical coastal and marine classification system. National parks, conservation and development. The role of protected areas in sustaining society.* Gland: IUCN.

Raz-Guzmán, A. and A.J. Sánchez. 1992. Quelonios: tortugas marinas – mamíferos marinos: grandes cetáceos: ballenas y rorcuales – mamíferos marinos: pequeños cetáceos: delfines, orcas, marsopas cachalotes – mamíferos marinos: sirenios y pinípedos: manatíes, focas y lobos marinos. In *Biología marina 2. flora y vertebrados. oceanografía. naturaleza. mapa IV. 9.4, atlas nacional de México.* México: Instituto de Geografía. UNAM. Sistemas de Información Geográfica, S.A.

Reyes Bonilla, H. and L.E. Calderon Aguilera. 1994. Parámetros poblacionales de Porites panamensis Verrill (Anthozoa: Scleractinia), en el arrecife de Cabo Pulmo, México. *Rev. Biol. Trop. Vol.* 42(1-2):121128.

Reyes Bonilla, H. and L.E. Calderon-Aguilera. 1999. Population density, distribution and consumption rates of three corallivores at Cabo Pulmo Reef, Gulf of California, Mexico. *Marine Ecology* 20 (3-4): 347-357.

Richards, R.A. and D.G. Deuel. 1987. Atlantic striped bass: stock status and the recreational fishery. *Mar. Fish. Rev.* 49(2): 58-66.

Roberts, C.M. 2002. Deep impact: the rising toll of fishing in the deep sea. *Trends in Ecology & Evolution* 17(5): 242-245.

Roberts, C.M., S. Andelman, G. Branch, R.H. Bustamante, J.C. Castilla, J. Dugan, B.S. Halpern, K.D. Lafferty, H. Leslie, J. Lubchenco, D. McArdle, H. Possingham, M. Ruckelshaus, R.R. Warner. 2003. Ecological criteria for evaluating candidate sites for marine reserves. *Ecological Applications* 13(1) Supplement: S199-S214.

Roberts, C.M., G. Branch, R.H. Bustamante, J.C. Castilla, J. Dugan, B.S. Halpern, K.D. Lafferty, H. Leslie, J. Lubchenco, D. McArdle, M. Ruckelshaus, R.R. Warner. 2003. Application of ecological criteria in selecting marine reserves and developing reserve networks. *Ecological Applications* 13(1) Supplement: S215-S228.

Robinson, C.L.K. and C.D. Levings. 1995. *An overview of habitat classification systems, ecological models and geographic information systems applied to shallow foreshore marine habitats. Canadian Manuscript Report of Fisheries and Aquatic Sciences 2322,* Ottawa: Department of Fisheries and Oceans.

Roden, G.I. 1972. Thermohaline structure and baroclinic flow across the Gulf of California entrance and the Revillagigedo Islands region. *Journal of Physical Oceanography* 2: 177-183.

Rodriguez Palacios, C. A., L.M. Mitchell Arana, G. Sandoval Diaz, P. Gomez and G. Green, 1988. Los moluscos de las Bahias de Huatulco y Puerto Angel, Oaxaca. Distribucion, diversidad y abundancia. *Universidad y Ciencia* 5(9): 85-94.

Roemmich, D. and J. McGowan. 1995. Climatic warming and the decline of zooplankton in the California Current. *Science* 267: 1324-1326.

Rosales, M.T.L., R.L. Escalona, R.M. Alarcon and V. Zamora. 1985. Organochlorine hydrocarbon residues in sediments of two different lagoons of northwest Mexico. *Bull. Environ. Contam. Toxicol.* 35: 322-330.

Rosales-Hoz, L. and A. Carranza-Edwards. 1990. Polymetallic nodule study from the oceanic area near Clarion Island, Mexico. *Marine Mining* 9:355-364.

Rosales-Hoz, L. and A. Carranza-Edwards. 1993. Geochemistry of deep-sea surface sediments from the Pacific manganese nodule province near Clarion Island, México. *Marine Georesources & Geotechnology* 11: 201-211.

Rosales-Hoz, L. and A. Carranza-Edwards. 2001. Geochemistry of deep-sea sediment cores and their relationship with polymetallic nodules from the north-eastern Pacific. *Marine and Freshwater Research* 52: 259-266.

Rosales-Hoz, L., A. Carranza-Edwards, S. Arias-Reynada and S. Santiago-Pérez. 1992. Distribución de metales en sedimentos recientes del sureste del Golfo de México. *An. Inst. Cienc. del Mar y Limnol., Univ. Nal. Autón. México* 19(2): 123-130.

Rosales-Hoz, L., A. Carranza-Edwards, C. Méndez-Jaime and M.A. Monreal-Gómez. 1999. Metals in shelf sediments and their association with continental discharges in a tropical zone. *Marine and Freshwater Research* 50: 189-196.

Rosales-Hoz, L., A. Carranza-Edwards, S. Santiago-Pérez, C. Méndez-Jaime and R. Doger-Badillo. 1994. Study of anthropogenically induced trace metals on the continental shelf in the southeastern part of the Gulf of Mexico. *Rev. Int. Cont. Amb.* 10 (1): 9-13.

Rowell, K., K.W. Flessa, D.L. Dettman, M.J. Román, L.H. Gerber and L.T. Findley. 2008. Diverting the Colorado River leads to a dramatic life history shift in an endangered marine fish. *Biological Conservation* Vol.141, No.4:1138-1148.

Salazar-Vallejo S.I. and N.E. González, ed. 1993. *Biodiversidad marina y costera de México.* México: Conabio y CIQRO.

Sánchez-Salazar, M.T. and A. Sánchez-Crispín. 1992. Producción pesquera y acuícola, 1989. In *Economía pesquera. pesca. mapa VI.5.1 atlas nacional de México.* México: Instituto de Geografía. UNAM. Sistemas de Información Geográfica, S.A.

Sánchez-Gil, P. and A. Yáñez-Arancibia. 1997. Grupos ecológicos funcionales y recursos pesqueros tropicales. *EPOMEX Serie Científica* 7: 357-389.

Santamaría-del-Angel E. and S. Alvarez-Borrego. 1994. Gulf of California biogeographic regions based on coastal zone color scanner imagery. *Journal of Geophysical Research* 99 (C4 April 15 1994): 7411 – 7421.

Schiff, K.C., M.J. Allen, E.Y. Zeng and S.M. Bay. 2000. Southern California Bight. In *Seas at the millenium: an environmental evaluation*, R.C. Shepherd, ed., Oxford: Pergamon Press.

Sherman, K and L.M. Alexander, ed. 1986. *Variability and management of large marine ecosystems. AAAS Selected Symposium 99*. Boulder, CO: Westview Press, Inc.

Sherman, K and A.M. Duda. 1999. An ecosystem approach to global assessment and management of coastal waters. Marine Ecology Progress Series. 190:271-287

Sherman, K. and G. Hempel (eds). 2009. *The UNEP Large Marine Ecosystem report: A perspective on changing conditions in LMEs of the world¹s regional seas.* UNEP Regional Seas Report and Studies No. 182. Nairobi, Kenya: United Nations Environment Programme.

Shluker, A.D. 2003. *State of Hawaii aquatic invasive species (AIS) management plan*. State of Hawaii Division of Aquatic Resources.

Shaw, R. 2000. Floyd wakes states to hazards of hog waste. *Environmental News Network* Sept 28, 2000.

Spalding, M.D., H.E. Fox, G.R. Allen, N. Davidson, Z.A. Ferdaña, M. Finlayson, B.S. Halpern, M.A. Jorge, A. Lombana, S.A. Lourie, K.D. Martin, E. McManus, J.Molnar, C.A. Recchia and J. Robertson. 2007. A bioregionalization of coastal and shelf areas. *Bioscience* 57(7): 573-583.

Springer, A.M. 1996. Prerecruit walleye pollock (*Theragra chalcogramma*) in seabird food webs of the Bering Sea.98-201. *U.S Dep. Commer. NOAA Tech. Rep. NMFS* 126: 198-201. RP-97-05.

Strub, P.T., HP Batchelder, and T J Weingartner. 2002. US GLOBEC Northeast Pacific Program: Overview. *Oceanography* 15(2) 2002: 30-35.

Sullivan Sealey, K and G. Bustamante. 1999. *Setting geographic priorities for marine conservation in Latin America and the Caribbean*. Arlington, Virginia: The Nature Conservancy.

Tapia-García, M. 1998. Ecology of demersal fish communities. Chap. 10. In *El Golfo de Tehuantepec: el ecosistema y sus recursos*, M. Tapia-García ed., 129-148. Iztapalapa, México: Universidad Autónoma Metropolitana-Iztapalapa, México.

Tapia García, M., ed. 1998. *El Golfo de Tehuantepec: el ecosistema y sus recursos*. Iztapalapa, México: Universidad Autónoma Metropolitana Iztapalapa.

Tapia-García, M. and M. C. García-Abad. 1998. Fish of shrimp bycatch and their potential as a resource in Oaxaca and Chiapas. Chap. 13. In *El Golfo de Tehuantepec: el ecosistema y sus recursos,* M. Tapia-García ed., 179-196. Iztapalapa, México: Universidad Autónoma Metropolitana-Iztapalapa.

Tapia-García M. and B. Gutiérrez Díaz. 1998. Fishing resources of the states of Oaxaca and Chiapas. Chap. 11. In *El Golfo de Tehuantepec: el ecosistema y sus recursos*, M. Tapia-García ed., 149-162. Iztapalapa, México: Universidad Autónoma Metropolitana-Iztapalapa, México.

Tapia-García M., E. Ramos-Santiago and A. Ayala-Cortés. 1998. The human activity and its impact on the coastal zone, with emphasis on the Isthmus of Tehuantepec. Chap. 15. In *El Golfo de Tehuantepec: el ecosistema y sus recursos*, M. Tapia-García ed., 209-228. Universidad Autónoma Metropolitana-Iztapalapa, México.

Tapia-García, M., F. Vázquez-Gutiérrez and A. Carranza-Edwards. 2001. Importancia de los vientos Tehuantepecanos en la dinámica ecológica del Golfo de Tehuantepec. *Memorias del IX congreso Latinoamericano e ibérico de meteorología, 7-11 de mayo, Buenos Aires, Argentina*, 6.A.5: 323-332.

TNC. 2000. *Identification of priority sites for conservation in the northern Gulf of México: an ecoregional plan*. Unpublished.The Nature Conservancy. October 2000.

Turgeon, D.D., R.G. Asch, B.D. Causey, R.E. Dodge, W. Jaap, K. Banks, J. Delaney, B.D. Keller, R. Speiler, C.A. Matos, J.R. Garcia, E. Diaz, D. Catanzaro, C.S. Rogers, Z. Hillis-Starr, R. Nemeth, M. Taylor, G.P. Schmahl, M.W. Miller, D.A. Gulko, J.E. Maragos, A.M. Friedlander, C.L. Hunter, R.S. Brainard, P. Craig, R.H.

Richond, G. Davis, J. Starmer, M. Trianni, P. Houk, C.E. Birkeland, A. Edward, Y. Golbou, J. Gutierrez, N. Idechong, G. Paulay, A. Tafileichig and N. Vander Velde. 2002. *The State of coral reef ecosystems of the United States and Pacific freely associated states: 2002*. Silver Spring, MD: National Oceanic and Atmospheric Administration/National Ocean Service/National Centers for Coastal Ocean Science.

Twichell, D.C., W.C. Schwab and N. Kenyon. 2005. Review of recent depositional processes on the Mississippi Fan, Gulf of Mexico. *Geological Society of America Abstracts with Programs* 36(5): 303.

Urbán R., J., L. Rojas-Bracho, M. Guerrero-Ruíz, A. Jaramillo-Legorreta and L. Findley. 2005. Cetacean diversity and conservation in the Gulf of California. In: *Biodiversity, Ecosystems, and Conservation in Northern Mexico.* J.-L. Cartron, G. Ceballos, and R. Felger, eds., 276-297. New York: Oxford University Press.

Vazquez Gutierrez, G. S. Lopez, A. Ramírez Alvarez, M. Turner Garces, A. F. Castillo and H. Alexander. 1998. Water chemistry. Chap. 4. In *El Golfo de Tehuantepec: el ecosistema y sus recursos,* M. Tapia-García ed., 35-50. Iztapalapa, México: Universidad Autónoma Metropolitana-Iztapalapa, México.

Wahl, T.R., K.H Morgan and K.Vermeer. 1993. Seabird distribution of British Columbia and Washington. In *Status, ecology and conservation of marine birds in the North Pacific*. K. Vermeer, K.Briggs, K. Morgan, D. Siegel-Causey ed., 39–47. Delta, B.C.: Canadian Wildlife Service Special Publication, Environment Canada.

Wiken, E.B. 1986. *Terrestrial ecozones of Canada. Ecological land classification series No. 19*. Ottawa: Environment Canada.

Wiken, E.B., D. Gauthier, I. Marshall, K. Lawton and H. Hirvonen. 1996. A perspective on Canada's ecosystems: an overview of the terrestrial and marine ecozones. Canadian Council on Ecological Areas (CCEA). Occasional Paper No. 14. Ottawa: CCEA.

Wiken, E.B., J. Robinson and L. Warren. 1998. Return to the sea: conservation of Canadian marine and freshwater ecosystems for wildlife In *Proceedings of a marine heritage conservation areas workshop, 3 April 1998. Waterloo, Ontario. Heritage Resources Centre, University of Waterloo* B.S. Iisaka, K. Van Osch and J. G. Nelson ed., 7 –18. Heritage Center Working Paper 14.

Wildlife Habitat Canada (WHC). 2001. The status of wildlife habitats in Canada's oceanic and coastal seascapes. Ottawa, ON: Wildlife Habitat Canada. Available: < http://www.whc.org/documents/FinalOceansLong VersionEnglish.pdf >.

Wilkinson, T., T. Agardy, S. Perry, L. Rojas, D. Hyrenbach, K. Morgan, D. Fraser, L. Janishevski, H. Herrmann and H. de la Cueva. 2004. Marine species of common conservation concern: protecting species at risk across international boundaries. In *Making ecosystem-based management work. Proceedings of the fifth international conference on science and the management of protected areas. Held at the University of Victoria, Victoria, B.C. May 11-16, 2003*.N.W.P Munro, P. Dearden, T.B. Herman, K. Beazley and S. Bondrup-Nielson, ed.Wolfville, NS: SAMPAA.

William, C.A. 1999. Rare and endangered marine invertebrates in British Columbia. In *Proceedings of a conference on the biology and management of species and habitats at risk, Kamloops, B.C. February 1999*, M. Darling ed., 57-65. Kamloops, BC: Ministry of Environment, Lands and Parks, Victoria, BC and University College of the Cariboo.

Williams, E. H., Jr., L. Bunkley-Williams, C.G. Lilyestrom, R.J. Larson, N.A. Engstrom, E.A.R. Ortiz-Corps and J.H. Timber. 2001. A population explosion of the rare tropical/subtropical purple sea mane, Drymonema dalmatinum, around Puerto Rico in the summer and fall of 1999. *Caribbean Journal of Science* 37(1-2): 127-130.

Yáñez-Arancibia, A. 1978. *Taxonomía, ecología y estructura de las comunidades de peces en lagunas costeras del Pacífico de México.* México: Centro Cienc. del Mar y Limnol. UNAM, Publ. Espec. 2.

Yáñez-Arancibia, A., ed. 1985. *Fish community ecology in estuaries and coastal lagoons: towards an ecosystem integration.* México, D.F.: ICML - UNAM, Editorial Universitaria.

Yáñez-Arancibia, A. 1987. Lagunas costeras y estuarios: cronología, criterios y conceptos para una clasificación ecológica de sistemas costeros. *Revta. Soc. Mex. Hist. Nat.* 39: 35-54.

Yáñez-Arancibia, A., ed. 1994. *Recursos faunísticos del litoral de la península de Yucatán. UAC-EPOMEX Serie Científica 2.* Campeche: UAC-EPOMEX.

Yáñez-Arancibia, A. 1999. Terms of references towards coastal management and sustainable development in Latin America: introduction to special issue on progress and experiences. *Ocean & Coastal Management* 42 (2.4): 77-104.

Yáñez-Arancibia, A. 2000. Coastal management in Latin America. Chap. 28. In *Seas at the millenium: an environmental evaluation,* C. Sheppard, ed., 457-466. Amsterdam: Pergamon/Elsevier Science Ltd.

Yáñez-Arancibia, A. 2002. Geomorphology and ecology of coastal zone in Middle America. In *Encyclopedia of coastal science,* M. Schwartz, ed. London, UK: Kluwer Academic Publs.

Yáñez-Arancibia, A. and A.L. Lara-Domínguez, ed. 1999. Mangrove ecosystems in tropical America. México, Costa Rica, Silver Spring: INECOL A.C., México UICN/ORMA Costa Rica, NOAA/NMFS Silver Spring, MD.

Yáñez-Arancibia, A., A.L. Lara-Domínguez, J.L. Rojas, D. Zárate Lomelí, G. J. Villalobos and P. Sánchez-Gil. 1999. Integrating science and management on coastal marine protected areas in the southern Gulf of Mexico. *Ocean & Coastal Management* 42 (2-4): 319-344.

Yáñez-Arancibia, A., P. Sánchez-Gil and A.L. Lara-Domínguez. 1991. Interacciones ecológicas estuario-mar: estructura funcional de bocas estuarinas y su efecto en la productividad del ecosistema. *Academia de Ciencias de Sao Paulo Brasil. Publ. ACIESP* 71(4): 49-83.

Yáñez-Arancibia, A., P. Sánchez-Gil and A.L. Lara-Dominguez. 1998. Functional groups and ecological biodiversity in Terminos Lagoon, *Mexico.Rev. Soc. Mex. Hist. Nat.* 49: 163-172.

Yáñez-Arancibia, A. and D. Zárate Lomelí, ed. 2002. *Términos de referencia para la gestión integrada de la zona costera del Golfo de México y Caribe. Panel MIZC-Golfo/Caribe, Xalapa 21-23 Nov. 2001.* México: INECOL A.C., Conanp-Semarnap, Conacyt Sisierra/Sigolfo.

Yáñez-Arancibia, A., D. Zárate Lomelí and V. Santiago Fandiño. 1996. La evaluación del impacto ambiental en la región del Gran Caribe. *Golfo de México contaminación e impacto ambiental: diagnóstico y tendencias, UAC-EPOMEX Serie Científica 5*: 587-604.

Zacharias, M.A., M.C. Morris and D.E. Howes 1999. Large scale characterisation of intertidal communities using a predictive model. *Journal of Experimental Marine Biology and Ecology* 239: 223–242.

Zárate Lomelí, D., G. Palacio, A. Yáñez-Arancibia, J.L. Rojas, A.L. Lara-Domínguez, G.J. Villalobos, J.F. Mas and A. Pérez. 1996. Remote sensing and GIS applied in the definition and zonification of the coastal areas in the southern Gulf of Mexico. In *Remote sensing for marine & coastal environments,* 625-631. Orlando: FL ERIM Publs.

Zárate Lomelí, D., J.L. Rojas Galaviz and T. Saavedra. 1996. La evaluación del impacto ambiental en México: recomendaciones para zonas costeras. *Golfo de México contaminación e impacto ambiental: diagnóstico y tendencias, UAC-EPOMEX Serie Científica 5*: 571-586.

Zárate Lomelí, D. T. Saavedra, J.L. Rojas, A. Yáñez-Arancibia and E. Rivera Arriaga.1999. Terms of reference towards an integrated management policy in the coastal zone of the Gulf of Mexico and the Caribbean. *Ocean & Coastal Management* 42 (2-4): 345-368.

Zwanenburg, K.C.T., D. Bowen, A. Bundy, K. Drinkwater, K. Frank, R. O'Boyle, D. Sameoto and M. Sinclair, 2002. Decadal changes in the Scotian Shelf large marine ecosystem. In *Large marine ecosystems of the North Atlantic—changing states and sustainability* K. Sherman and H.R. Skjoldal, ed., 105-150. Elsevier.

Mangrove swamp on the Pacific coast. La Encrucijada Biosphere Reserve, Chiapas.   *Photo:* Patricio Robles Gil

Serendipitous photo of the shy and rare vaquita (*Phocoena sinus*), a species of porpoise endemic to the northern extreme of the Gulf of California, Mexico. It is the most critically endangered of all cetaceans, and also has the most limited distribution.   *Photo:* Chris Johnson/EarthOCEAN

Web sites

http://bonita.mbnms.nos.noaa.gov/sitechar/cold3.html

http://cbl.umces.edu

http://oceanexplorer.noaa.gov/explorations/islands01/log/sep27/sep27.html

http://radlab.soest.hawaii.edu/atlas/

http://topex.ucsd.edu/marine_topo/mar_topo.html

http://wwf.ca/AboutWWF/WhatWeDo/TheNatureAudit/

http://www.afsc.noaa.gov/Quarterly/ond2003/divrptsNMML3.htm

http://www.aquatic.uoguelph.ca/index.htm

http://www.baruch.sc.edu

http://www.ccea.org

http://www.chesapeakebay.net

http://www.cisti.nrc.ca

http://www.csc.noaa.gov/crs/bhm/online.html

http://www.epa.gov/mrlcpage

http://www.epa.gov/msbasin/

http://www.fishbase.org/home.htm

http://www.gbif.org/

http://www.gov.mb.ca/natres/watres/wrb_main.html

http://www.imo.org

http://www.invasivespeciesinfo.gov/profiles/rapawhelk.shtml

http://www.nmfs.noaa.gov/sfa/sfweb/index.htm

http://www.oar.noaa.gov/oceans

http://www.oceanexplorer.noaa.gov

http://www.ovi.ca/status.html

http://www.pc.gc.ca/progs/amnc-nmca/plan/gloss/polynya_e.asp

http://www.pmel.noaa.gov/

http://www.pmel.noaa.gov/sebscc/concept/index.html

http://www.slv2000.qc.ec.gc.ca

http://www.soundkeeper.org

http://nsidc.org/arcticmet/patterns/arctic_oscillation.html

http://jisao.washington.edu/pdo/

http://www.uaf.edu/ces/aiswg/pdf-documents/AISWGminutes-12-11-06.pdf

http://woodshole.er.usgs.gov/project-pages/stellwagen/didemnum/images/pdf/news/pnwscuba.pdf

California sea lions find a place to rest and breed on the rocky shoreline of San Pedro Mártir island, in the Gulf of California.   *Photo:* Patricio Robles Gil

Common Name in English (Some species have more than one common name)	At risk	Nom latin	Nombre común en español (Algunas especies tienen más de un nombre común)	Nom commun en français (certaines espèces ont plus d'un nom commun)	Featured in Ecoregion	Photo (page)
Mammals			**Mamíferos**	**Mammifères**		
Bear, Kodiak		*Ursus arctos middendorffi*	Oso Kodiak	Kodiak de l'Alaska	22 23	*128–129*
Bear, Polar	▲	*Ursus maritimus*	Oso polar	Ours blanc	1 2 3 4 5 6	*xvi, 22–23*
Dolphin, Atlantic Spotted		*Stenella frontalis*	Delfín manchado del Atlántico	Dauphin tâcheté de l'Atlantique	8 9 10 11 12 13 14 15	*xviii*
Dolphin, Atlantic White-sided		*Lagenorhynchus acutus*	Delfín de costados blancos del Atlántico	Dauphin à flancs blancs de l'Atlantique	9	
Dolphin, Bottlenose		*Tursiops truncatus*	Delfín mular o nariz de botella	Dauphin à gros nez	9 19	*108*
Dolphin, Common; Short-beaked Common Dolphin		*Delphinus delphis*	Delfín común (de rostro corto)	Dauphin commun	9	
Dolphin, Long-beaked Common		*Delphinus capensis*	Delfín común de rostro largo	Dauphin commun à long bec	18	
Dolphin, Pacific White-sided		*Lagenorhynchus obliquidens*	Delfín de costados blancos del Pacífico	Dauphin à flancs blancs du Pacifique.	22	
Dolphin, Risso's		*Grampus griseus*	Delfín de Risso	Dauphin de Risso	9	
Elephant Seal, Northern		*Mirounga angustirostris*	Elefante marino del norte	Éléphant de mer boréal	19 20 23	
Fishing Bat, Gulf; Fish-eating Bat		*Myotis vivesi*	Murciélago pescador	*Myotis vivesi*†	18 §	
Fox, Arctic		*Vulpes lagopus*	Zorro ártico	Renard arctique	22 *	
Fur Seal, Guadalupe	▲	*Arctocephalus townsendi*	Lobo fino de Guadalupe o foca de Guadalupe	Otarie à fourrure de Townsend	19	
Manatee, Florida	▲	*Trichechus manatus latirostris*	Manatí de Florida	Lamantin de Floride	12 14 15	*67*
Manatee, West Indian	▲	*Trichechus manatus*	Manatí o vaca marina	Lamantin des Caraïbes	11 12 14 15	*xi*
Monk Seal, Hawaiian	▲	*Monachus schauinslandi*	Foca monje de Hawai	Phoque moine d'Hawaï	24 §	*141*
Narwhal	▲	*Monodon monoceros*	Narval o ballena unicornio	Narval	3 4 5 6	*41*
Otter, Sea	▲	*Enhydra lutris*	Nutria marina (del sur)	Loutre de mer	19 20 21 22 23	
Pilot Whale, Long-Finned		*Globicephala melas*	Ballena piloto	Globicéphale noir	6	
Porpoise, Dall's		*Phocoenoides dalli*	Marsopa de Dall	Marsouin de Dall	20 22	
Porpoise, Harbor		*Phocoena phocoena*	Marsopa común	Marsouin commun	7 22	
Porpoise, Vaquita	▲	*Phocoena sinus*	Vaquita	Marsouin du golfe de Californie	18 §	*162*
Rat, Norway		*Rattus norvegicus*	Rata noruega, café o de alcantarilla	Rat surmulot	22 *	
Sea Lion, California		*Zalophus californianus*	Lobo marino de California	Otarie de Californie	18 19 21	*164*
Sea Lion, Steller	▲	*Eumetopias jubatus*	Lobo marino de Steller	Otarie de Steller	1 2 20 21 22 23	
Seal, Bearded		*Erignathus barbatus*	Foca barbada	Phoque barbu	2 3 6 7	
Seal, Gray		*Halichoerus grypus*	Foca gris	Phoque gris	7 9	
Seal, Harbor		*Phoca vitulina*	Foca común	Phoque commun	1 3 4 5 6 7 20 21 22	
Seal, Harp		*Phoca groenlandica*	Foca de Groenlandia	Phoque du Groenland	3 4 5 6 7	
Seal, Hooded		*Cystophora cristata*	Foca de capuchón	Phoque à capuchon	6 7	
Seal, Northern Fur		*Callorhinus ursinus*	Lobo fino del norte o foca de Alaska	Otarie à fourrure	1 22 23	*136*
Seal, Ribbon		*Histriophoca fasciata*	Foca franjeada	Phoque à bandes	1	

▲ species at risk * introduced and invasive species § endemic species † species for which no common name could be found in the literature consulted.

Notes : For each species featured throughout the book as occurring in one or more ecoregions, it should be understood that this is in no way intended to restrict its distribution to just the regions mentioned: marine species often range freely over considerable distances or are even highly migratory in their feeding/reproductive habits.

A species whose range of distribution is far from a region where a given language is spoken may not necessarily have a common name in that language. In such cases, the scientific (Latin) name, a vulgarization of it, or a variant of the name in the local language is usually taken into English as the common name.

Note that some species may have more than one common name. The present list should not be taken as definitive and anyone having corrections or additions to suggest is cordially invited to communicate them to the CEC Secretariat.

Common Name in English (Some species have more than one common name)	At risk	Nom latin	Nombre común en español (Algunas especies tienen más de un nombre común)	Nom commun en français (certaines espèces ont plus d'un nom commun)	Featured in Ecoregion	Photo (page)
Seal, Ringed		*Phoca hispida*	Foca anillada	Phoque annelé	2 3 4 5 6 7	
Seal, Spotted		*Phoca largha*	Foca manchada	Phoque circumpolaire	1	
Walrus, Atlantic	▲	*Odobenus rosmarus rosmarus*	Morsa del Atlántico	Morse de l'Atlantique	5 6	
Walrus, Pacific	▲	*Odobenus rosmarus divergens*	Morsa del Pacífico	Morse du Pacifique	1 2 3	*14–15*
Whale, Beluga	▲	*Delphinapterus leucas*	Beluga o ballena blanca	Béluga	2 3 4 5 6 7 22	*36–37*
Whale, Blue	▲	*Balaenoptera musculus*	Ballena azul	Rorqual bleu	1 6 7 18 19 20 21 22 23	*104, 150–151*
Whale, Bowhead	▲	*Balaena mysticetus*	Ballena de Groenlandia	Baleine boréale	1 2 4 5 6 7 22 23	*19*
Whale, Bryde's	▲	*Balaenoptera edeni*	Ballena de Bryde o rorcual tropical	Balénoptère boréal de Bryde ou baleine de Bryde	18 19	
Whale, Fin	▲	*Balaenoptera physalus*	Ballena de aleta o rorcual común	Rorqual commun	1 6 7 8 9 10 11 18 19 20 21	
Whale, Gray	▲	*Eschrichtius robustus*	Ballena gris	Baleine grise	1 2 18 19 20 21 22	*110*
Whale, Humpback	▲	*Megaptera novaeangliae*	Ballena jorobada	Rorqual à bosse	6 7 9 10 18 20 21 22 23 24	*130–131, 140*
Whale, Killer; Orca	▲	*Orcinus orca*	Orca	Épaulard ou orque	6 21 22	*120–121*
Whale, Minke	▲	*Balaenoptera acutorostrata*	Ballena de minke, rorcual aliblanco o rorcual menor	Petit rorqual	6 7 18 19 21 22	
Whale, North Atlantic Right	▲	*Eubalaena glacialis*	Ballena franca boreal o franca del Atlántico norte	Baleine noire de l'Atlantique Nord	6 7 8 9 10 11	*51*
Whale, North Pacific Right	▲	*Eubalaena japonica*	Ballena franca del Pacífico norte	Baleine noire du Pacifique Nord	1 20 21 22 23	
Whale, Northern Bottlenose		*Hyperoodon ampullatus*	Ballena nariz de botella	Baleine à bec commune	6	
Whale, Sei	▲	*Balaenoptera borealis*	Ballena sei o boreal	Rorqual boréal	6 7 18 19 22 23	
Whale, Sperm	▲	*Physeter macrocephalus*	Cachalote	Grand cachalot	6 7 9 10 18 20 21 22 23	*55*
Birds			**Aves**	**Oiseaux**		
Albatross, Laysan		*Phoebastria immutabilis*	Albatros de Laysan	Albatros de Laysan	20 23	
Albatross, Short-tailed	▲	**Phoebastria albatrus**	**Albatros de cola corta**	**Albatros à queue courte**	1 19 20 23 24	
Alcids and Auks		**Family: *Alcidae***	**Álcidos (alcas, alcitas y alcuelas)**	**Alcidés**	4 7 9	
Auklet, Cassin's		**Ptychoramphus aleuticus**	**Alcuela o alcita oscura**	**Starique de Cassin**	1 22	
Auklet, Crested		**Aethia cristatella**	**Alcuela crestada**	**Starique cristatelle**	1 23	*12*
Auklet, Least		**Aethia pusilla**	**Alcita pequeña**	**Starique minuscule**	1 23	*12*
Auklet, Whiskered		**Aethia pygmaea**	**Alcuela enana**	**Starique pygmée**	1 23	
Booby, Blue-footed		*Sula nebouxii*	**Bobo de patas azules**	**Fou à pieds bleus**	18	*103*
Booby, Masked		*Sula dactylatra*	**Bobo enmascarado**	**Fou masqué**	14	*76*
Brant		*Branta bernicla*	**Ganso de collar**	**Bernache cravant**	2 5 22	
Cormorant, Double-crested		*Phalacrocorax auritus*	**Cormorán orejudo**	**Cormoran à aigrettes**	7	
Cormorant, Neotropical		*Phalacrocorax brasilianus*	**Cormorán neotropical u oliváceo**	**Cormoran vigua**	15	*83*
Dovekie		*Alle alle*	**Mérgulo atlántico**	**Mergule nain**	6	
Duck, Long-tailed; Oldsquaw		*Clangula hyemalis*	**Pato cola larga**	**Harelde kakawi**	2 5	
Eagle, Bald	▲	*Haliaeetus leucocephalus*	**Águila cabeza blanca**	**Pygargue à tête blanche**	12	
Eider, Common		*Somateria mollissima*	**Eider común**	**Eider à duvet**	1 5 7	
Eider, King	▲	*Somateria spectabilis*	**Eider real**	**Eider à tête grise**	1 2 6	*35*
Eider, Steller's	▲	*Polysticta stelleri*	**Eider de Steller**	**Eider de Steller**	1	
Falcon, Peregrine	▲	*Falco peregrinus*	**Halcón peregrino**	**Faucon pèlerin**	5	

▲ species at risk   * introduced and invasive species   § endemic species   † species for which no common name could be found in the literature consulted.

Common Name in English (Some species have more than one common name)	At risk	Nom latin	Nombre común en español (Algunas especies tienen más de un nombre común)	Nom commun en français (certaines espèces ont plus d'un nom commun)	Featured in Ecoregion	Photo (page)
Flamingo		*Phoenicopterus ruber*	Flamenco rosado	Flamand rose	14	77
Fulmar, Northern		*Fulmarus glacialis*	Fulmar del norte	Fulmar boréal	4 5 6 7 9 23	
Gannet, Northern		*Morus bassanus*	Bobo norteño	Fou de Bassan	6 7 9	40
Godwit, Hudsonian		*Limosa haemastica*	Picopando ornamentado	Barge hudsonienne	5	
Goose, Canada		*Branta canadensis*	Ganso canadiense	Bernache du Canada	2 4 5 22 23	
Goose, Greater White-fronted		*Anser albifrons*	Ganso careto mayor	Oie rieuse	2 5 22	
Goose, Grey		*Anser spp.*	Ganso cenizo o ánsar	Oies	2 4 22	
Goose, Ross's		*Chen rossii*	Ganso de Ross	Oie de Ross	2 5	
Goose, Snow		*Chen caerulescens*	Ganso nevado	Oie des neiges	2 4 5	
Guillemot, Black; Tystie		*Cepphus grylle*	Arao aliblanco	Guillemot à miroir	2	
Guillemot, Pigeon		*Cepphus columba*	Arao pichón	Guillemot colombin	23	
Gull, California		*Larus californicus*	Gaviota californiana	Goéland de Californie	20	
Gull, Glaucous		*Larus hyperboreus*	Gaviota blanca	Goéland bourgmestre	5	
Gull, Great Black-backed		*Larus marinus*	Gaviota sombría mayor o gavión atlántico	Goéland marin	7	
Gull, Heermann's		*Larus heermanni*	Gaviota ploma	Goéland de Heermann	18	
Gull, Iceland		*Larus glaucoides*	Gaviota de Islandia	Goéland arctique	5 6 7	
Gull, Ivory		*Pagophila eburnea*	Gaviota marfil	Mouette blanche	3 5 6	
Gull, Sabine's		*Xema sabini*	Gaviota cola hendida	Mouette de Sabine	5	
Gull, Yellow-footed		*Larus livens*	Gaviota pata amarilla	Goéland de Cortez	18	
Jaegers, Skuas		*Stercorarius spp.*	Págalos o salteadores	Labbes	4	
Kittiwake, Black-legged		*Rissa tridactyla*	Gaviota pata negra	Mouette tridactyle	1 5 7	
Kittiwake, Red-legged	▲	*Rissa brevirostris*	Gaviota de pico corto	Mouette des brumes	1	13
Least Tern, California	▲	*Sternula antillarum*	Gallito californiano	Petite sterne	19 20	
Loon, Yellow-billed; White-billed Diver		*Gavia adamsii*	Colimbo de Adams	Plongeon à bec blanc	5	
Loons		*Gavia spp.*	Colimbos	Plongeons	2 4 5 22	
Merganser, Common		*Mergus merganser*	Mergo mayor o común	Grand harle	2	
Merganser, Red-breasted		*Mergus serrator*	Mergo copetón	Harle huppé	2	
Murre, Common; Common Guillemot		*Uria aalge*	Arao común	Guillemot marmette	1 6 7 20 21 22 23	137
Murre, Thick-billed; Brunnich's Guillemot		*Uria lomvia*	Arao de Brünnich	Guillemot de Brünnich	2 5 6 7	18
Murrelet, Ancient		*Synthliboramphus antiquus*	Mérgulo antiguo	Guillemot à cou blanc	1	
Murrelet, Marbled	▲	*Brachyramphus marmoratus*	Mérgulo marmoleado	Guillemot marbré	1 20 21	
Murrelet, Xantus'	▲	*Synthliboramphus hypoleucus*	Mérgulo de Xantus	Guillemot de Xantus	19	
Pelican, Brown	▲	*Pelecanus occidentalis* [California subspecies: *Pelecanus occidentalis californicus*)]	Pelícano pardo	Pélican brun	18 20	114
Pelican, White		*Pelecanus erythrorhynchos*	Pelícano blanco	Pélican d'Amérique	13	73
Petrel, Hawaiian Dark-rumped	▲	*Pterodroma sandwichensis*	Petrel hawaiano	Pétrel des Hawaï	24	
Phalarope, Red-necked		*Phalaropus lobatus*	Falaropo cuello rojo o picofino	Phalarope à bec étroit	2	
Plover, Piping	▲	*Charadrius melodus*	Chorlo chiflador	Pluvier siffleur	12	
Puffin, Atlantic		*Fratercula arctica*	Frailecillo común o del Atlántico	Macareux moine	6 7	45

▲ species at risk   * introduced and invasive species   § endemic species   † species for which no common name could be found in the literature consulted.

Common Name in English (Some species have more than one common name)	At risk	Nom latin	Nombre común en español (Algunas especies tienen más de un nombre común)	Nom commun en français (certaines espèces ont plus d'un nom commun)	Featured in Ecoregion	Photo (page)
Puffin, Tufted		*Fratercula cirrhata*	Frailecillo de cola grande	Macareux huppé	22 23	135
Razorbill		*Alca torda*	Alca común	Petit pingouin	6	
Sandpiper, Semipalmated		*Calidris pusilla*	Playero semipalmeado	Bécasseau semipalmé		179
Scaup, Greater and Lesser		*Aythya marila, A. affinis*	Patos boludos, mayor y menor	Fuligule milouinan et petit fuligule	2	
Scoters		*Melanitta spp.*	Negretas	Macreuses	2 21	
Seaside Sparrow, Cape Sable	▲	*Ammodramus maritimus mirabilis*	Gorrión marino del cabo de Sable	Bruant maritime de Cap de Sable	12	
Shearwater, Greater		*Puffinus gravis*	Pardela mayor	Puffin majeur	6 7	
Shearwater, Newell's	▲	*Puffinus auricularis newellii*	Pardela de Newell	Puffin de Newell	24	
Shearwater, North Atlantic Little		*Puffinus baroli*	Pardela chica (del Atlántico norte)	Petit puffin de l'Atlantique Nord	6 7	
Shearwater, Pink-footed	▲	*Puffinus creatopus*	Pardela pata rosada	Puffin à pieds roses	1 19 20 21	123
Shearwater, Sooty		*Puffinus griseus*	Pardela sombría o gris	Puffin fuligineux	6 7	
Snow Goose, Greater		*Chen caerulescens atlantica*	Ganso nevado o blanco mayor	Grande oie des neiges	2 4 5	
Spoonbilll, Roseate		*Platalea ajaja* or *Ajaia ajaja*	Espátula rosada	Spatule rosée	16	87
Storm-Petrel, Ashy		*Oceanodroma homochroa*	Paíño cenizo	Océanite cendré	20	
Storm-Petrel, Band-Rumped; Madeiran Storm-Petrel	▲	*Oceanodroma castro*	Paíño de Madeira	Océanite de Castro	24	
Storm-Petrel, Leach's		*Oceanodroma leucorhoa*	Paíño de Leach	Océanite cul-blanc	7 9	
Swan, Tundra		*Cygnus columbianus*	Cisne de la tundra	Cygne siffleur	5 22	29
Tern, Arctic		*Sterna paradisaea*	Charrán ártico	Sterne arctique	4 5 7	28
Tern, Roseate	▲	*Sterna dougallii*	Charrán rosado	Sterne de Dougall	12 15	
Tern, Royal		*Thalasseus maxima*	Charrán real	Sterne royale	18	
Whimbrel		*Numenius phaeopus*	Zarapito trinador o picopando canelo	Courlis corlieu	5	
Marine Turtles			**Tortugas marinas**	**Tortues de mer**		
Sea Turtle, Leatherback	▲	*Dermochelys coriacea*	Tortuga laúd	Tortue luth	6 7 8 9 10 11 14 15 16 17 18 20 21 23 24	59
Sea Turtle, Loggerhead	▲	*Caretta caretta*	Caguama	Caouanne	9 11 12 13 14 15 16 17 18 20	
Turtle, Green; East Pacific Green, Or Atlantic Green	▲	*Chelonia mydas*	Tortuga verde, verde del Pacífico oriental o verde del Atlántico	Tortue verte	11 12 13 14 15 16 17 18 24	
Turtle, Hawksbill	▲	*Eretmochelys imbricata*	Tortuga de carey	Tortue imbriquée	11 12 14 15 24	63
Turtle, Kemp's Ridley	▲	*Lepidochelys kempii*	Tortuga lora	Tortue de Kemp	12 13 14	72
Turtle, Olive Ridley; Pacific Ridley Turtle	▲	*Lepidochelys olivacea*	Tortuga golfina, olivacea o escamosa del Pacífico	Tortue olivâtre	16 17 18	94
Alligators, Crocodiles, and Turtles			**Cocodrilos y tortugas**	**Crocodiles et tortues**		
Alligator, American	▲	*Alligator mississipiensis*	Caimán o aligátor americano	Alligator d'Amérique	12	
Crocodile, American	▲	*Crocodylus acutus*	Cocodrilo americano	Crocodile américian	12	
Terrapin, Diamondback	▲	*Malaclemys terrapin*	Tortuga de dorso diamantino	Tortue à dos diamanté	8 11 13	
Fish			**Peces**	**Poissons**		
Amberjack, Greater		*Seriola dumerili*	Medregal coronado	Sériole	12 14	142–143
Anchovies		*Anchoa spp., Anchovia spp., Cetengraulis spp.,* and *Engraulis spp.*	Anchovetas	Anchois	18	
Anchovy, Northern; Californian Anchoveta		*Engraulis mordax*	Anchoveta norteña	Anchois du Pacifique	18	
Angelfish, French		*Pomacanthus paru*	Pez ángel francés o gallineta negra	Poisson-ange français	12	82

▲ species at risk   * introduced and invasive species   § endemic species   † species for which no common name could be found in the literature consulted.

Common Name in English (Some species have more than one common name)	At risk	Nom latin	Nombre común en español (Algunas especies tienen más de un nombre común)	Nom commun en français (certaines espèces ont plus d'un nom commun)	Featured in Ecoregion	Photo (page)
Barred Pargo, Mexican; Barred Snapper		*Hoplopagrus guentheri*	Pargo coconaco	Vivaneau mexicain	18	
Bass, Parrot Sand		*Paralabrax loro*	Cabrilla loro o cachete amarillo	Serran perroquet	18	
Bass, Spotted Sand		*Paralabrax maculatofasciatus*	Cabrilla de roca o arenera	Serran de roche	18 19 20	
Bass, Striped		*Morone saxatilis*	Lubina estriada	Bar d'Amérique	8 11	
Blenny, Bay		*Hypsoblennius gentilis*	Borracho o trambollito de bahía	*Hypsoblennius gentilis*†	18 19 20	
Blenny, Fourline Snake		*Eumesogrammus praecisus*	*Eumesogrammus praecisus*†	Blennie quatre-lignes atlantique	5	
Blenny, Gulf Signal		*Emblemaria hypacanthus*	Trambollito vela	*Emblemaria hypacanthus*†	18 §	
Blenny, Sonora		*Malacoctenus gigas*	Trambollo de Sonora	Blennie de Sonora	18 §	
Bluefish		*Pomatomus saltatrix*	Anchoa	Tassergal	11 12	
Bonefish		*Albula vulpes*	Macabí o lisa francesa	Albula	12	
Bonito, Pacific		*Sarda chiliensis lineolata*	Bonita chilena	Bonite du Pacifique	19 21	
Butterflyfish		Family: *Chaetodontidae*	Peces mariposa	Chitons	10	
Butterflyfish, Threebanded		*Chaetodon humeralis*	Pez mariposa de tres bandas	Papillon à trois bandes	17	
Cabrilla, Flag		*Epinephelus labriformis*	Cabrilla piedrera	Mérou étoile	18	
Cabrilla, Spotted		*Epinephelus analogus*	Cabrilla pinta	Mérou cabrilla	18	
Capelin		*Mallotus villosus*	Capelán	Capelan	2 5 6	
Char, Arctic		*Salvelinus alpinus*	Trucha ártica o salvelino alpino	Omble chevalier	3 4 5 6	
Clingfish, Cortez		*Tomicodon boehlkei*	Chupapiedras de Cortés	*Tomicodon boehlkei*†	18 §	
Clingfish, Mexican		*Gobiesox mexicanus*	Cucharita mexicana	Gobie du Mexique	16 §	
Cod, Arctic		*Boreogadus saida*	Bacalao polar	Morue polaire ou saïda franc	2 3 4 5 6	25
Cod, Atlantic	▲	*Gadus morhua*	Bacalao del Atlántico	Morue franche	3 6 7 8 9	
Cod, Greenland; Rock Cod; Ogac		*Gadus ogac*	Bacalao de Groenlandia	Ogac	2 3 4 5 6	
Cod, Pacific		*Gadus macrocephalus*	Bacalao del Pacífico	Morue du Pacifique	1 22 23	
Cod, Polar		*Arctogadus glacialis*	Bacalao del Àrtico	Saïda imberbe	2 3 4	25
Coney, Gulf		*Epinephelus acanthistius*	Baqueta	Mérou coq	18	
Corvina, Gulf; Gulf Weakfish	▲	*Cynoscion othonopterus*	Corvina golfina	Acoupa du golfe	18 §	
Corvina, Orangemouth; Orangemouth Weakfish		*Cynoscion xanthulus*	Corvina boca amarilla	Acoupa à gueule jaune	18	
Corvina, Shortfin; Shortfin Weakfish		*Cynoscion parvipinnis*	Corvina azul	Acoupa magdalène	18	
Corvina, Striped		*Cynoscion reticulates*	Corvina rayada	Acoupa rayée	18	
Cowcod (Rockfish)	▲	*Sebastes levis*	Rocote vaquilla (pez roca)	*Sebastes levis*†	19 20 21	
Croaker, Atlantic		*Micropogonias undulatus*	Gurrubata	Tambour brésilien	10	
Croaker, Bigeye; Gulf Croaker		*Micropogonias megalops*	Chano norteño	Tambour à grand œil	18 §	
Croaker, Spot		*Leiostomus xanthurus*	Croca	Tambour croca	13	
Damselfish, Acapulco; Acapulco Major		*Stegastes acapulcoensis*	Jaqueta o damisela Acapulco	Chauffet Acapulco	17	
Damselfish, Cortez		*Stegastes rectifraenum*	Damisela o jaqueta de Cortés	*Stegastes rectifraenum*†	18 §	
Damselfish, Garibaldi		*Hypsypops rubicundus*	Garibaldi	Demoiselle Garibaldi	19	111
Damselfish, Giant		*Microspathodon dorsalis*	Castañuela gigante	Chauffet ayanque	17	
Dolphinfish; Dorado		*Coryphaena hippurus*	Dorado	Coryphène commune	10 18	

▲ species at risk   ∗ introduced and invasive species   § endemic species   † species for which no common name could be found in the literature consulted.

Common Name in English (Some species have more than one common name)	At risk	Nom latin	Nombre común en español (Algunas especies tienen más de un nombre común)	Nom commun en français (certaines espèces ont plus d'un nom commun)	Featured in Ecoregion	Photo (page)
Eel, green moray		*Gymnothorax castaneus*	Morena verde	Murène verte		*xiv–xv*
Eelblenny, Slender		*Lumpenus fabricii*	*Lumpenus fabricii* †	Lompénie élancée	5	
Eelpout, Arctic		*Lycodes reticulatus*	*Lycodes reticulatus* †	Lycode arctique	3 4	
Flounder, Arrowtooth		*Atheresthes stomias*	Halibut del Pacífico	Plie à grande bouche	22 23	
Flounder, Dappled		*Paralichthys woolmani*	Lenguado huarache	Cardeau huarache	18	
Flounder, Georges Bank Yellowtail		*Limanda ferruginea*	Platija amarilla del banco de Georges	Limande à queue jaune	7	
Flounder, Northern or Arctic		*Pleuronectes glacialis*	Platija del ártico o solla	Plie arctique	6	
Flounder, Oval		*Syacium ovale*	Lenguado ovalado	Hémirhombe oval	17	
Goatfish		*Pseudupeneus maculates; Mulloidichthys martinicus* (Family: *Mullidae*)	Múlidos	Rougets	10	
Gobies		Family: *Gobiidae*	Gobios	Gobies	15 §	
Goby, Bluebanded		*Lythrypnus dalli*	Gobio bonito	Gobie de Catalina	18 19 20	
Goby, Guaymas		*Quietula guaymasiae*	Gobio guaymense	*Quietula guaymasiae* †	18 §	
Gobies, Hawaiian	▲	*Bathygobius spp.*	Gobio hawaiano	*Bathygobius spp.* †	24 §	
Goby, Shadow; American Shadow Goby		*Quietula y-cauda*	Gobio sombreado	*Quietula y-cauda* †	18 19 20	
Goby, Slow		*Aruma histrio*	Gobio lento	*Aruma histrio* †	18 §	
Groupers	▲	*Mycteroperca spp.* and *Epinephelus spp.* (Family: *Serranidae*)	Meros, garropas y cabrillas (serránidos)	Mérous	10 13 14 18	
Grouper, Broomtail		*Mycteroperca xenarcha*	Cabrilla plomuda	Badèche balai	18	
Grouper, Goliath		*Epinephelus itajara*	Mero guasa o cherna	Mérou géant	10 13 14 18	
Grouper, Gulf		*Mycteroperca jordani*	Baya o garropa del golfo	Mérou golfe	18	
Grouper, Leopard		*Mycteroperca rosacea*	Cabrilla sardinera	Mérou léopard	18 §	
Grouper, Nassau	▲	*Epinephelus striatus*	Cherna criolla	Mérou de Nassau	15	
Grouper, Red		*Epinephelus morio*	Mero yucateco o rojo	Mérou rouge	14	
Grouper, Sawtail		*Mycteroperca prionura*	Cabrilla o garropa aserrada	Mérou scie-queue	18 §	
Grouper, Star-Studded		*Epinephelus niphobles*	Baqueta ploma o mero manchado	Mérou tacheté	18	
Grunion, Gulf		*Leuresthes sardina*	Gruñón o pejerrey sardina	Athérine mexicaine	18 §	
Grunt, Burrito		*Anisotremus interruptus*	Burrito	Lippu bourricot	17	
Grunt, Spottail		*Haemulon maculicauda*	Roncador esmeralda	Gorette à queue tachetée	16	88–89
Guitarfish, Speckled		*Rhinobatos glaucostigma*	Pez guitarra punteada	Poisson-guitare marbré	17	
Haddock		*Melanogrammus aeglefinus*	Eglefino	Églefin	6 7 8	
Hake, Pacific	▲	*Merluccius productus*	Merluza del Pacífico	Merlu du Pacifique nord	19 20 21	
Hake, Silver; Atlantic Hake, New England Hake	▲	*Merluccius bilinearis*	Merluza plateada	Merlu argenté	6 9 10	
Halibut, Atlantic		*Hippoglossus hippoglossus*	Lenguado del Atlántico	Flétan de l'Atlantique	6	
Halibut, Cortez		*Paralichthys aestuarius*	Lenguado alabato o de Cortés	Cardeau alabate	18	
Halibut, Greenland; Greenland Turbot		*Reinhardtius hippoglossoides*	Platija de Groenlandia	Flétan du Groenland	6 7 23	
Halibut, Pacific		*Hippoglossus stenolepis*	Lenguado del Pacífico	Flétan du Pacifique	1 22 23	
Herring, Atlantic		*Clupea harengus*	Arenque del Atlántico	Hareng atlantique	6 7 8 9	
Herring, Graceful; Graceful Piquitinga		*Lile gracilis*	Charal o sardinita agua dulce	Piquitingue du Pacifique	16 §	

▲ species at risk * introduced and invasive species § endemic species † species for which no common name could be found in the literature consulted.

Common Name in English (Some species have more than one common name)	At risk	Nom latin	Nombre común en español (Algunas especies tienen más de un nombre común)	Nom commun en français (certaines espèces ont plus d'un nom commun)	Featured in Ecoregion	Photo (page)
Herring, Pacific		*Clupea pallasii pallasii*	Arenque del Pacífico	Hareng du Pacifique	2 21 22	
Herring, Round		*Etrumeus teres*	Sardina japonesa	Shadine	6 7 18	
Hind, Speckled; Calico Grouper	▲	*Epinephelus drummondhayi*	Mero pintarroja o pintado	Mérou grivel	11 12 13 14	
Houndsharks; Smoothhounds	▲	*Mustelus spp.*	Cazones	Émissoles	18	
Jacks		Family: *Carangidae*	Jureles y pámpanos (carángidos)		14 18 24	
Jack, Almaco		*Seriola rivoliana*	Medregal limón	Sériole limon	18	
Jack, Black		*Caranx lugubris*	Jurel negro	Carangue noire	18 24	
Jack, Green		*Caranx caballus*	Cocinero, jurel dorado o jurel bonito	Carangue verte	18 24	
Jack, Pacific Crevalle		*Caranx caninus*	Jurel toro	Grande carangue du Pacifique	18 24	
Jack, Yellowtail; Yellowtail Amberjack		*Seriola lalandi*	Jurel aleta amarilla	Sériole chicard	18	
Lingcod		*Ophiodon elongatus*	Bacalao largo, bacalao ling o lorcha	Morue-lingue	20 21	
Lizardfish, Shorthead; Lance Lizardfish		*Synodus scituliceps*	Chile arpón	Anoli liguise	17	
Lionfish, Red		*Pterois volitans*	Pez león común	Rascasse volante	11 * 12 *	
Mackerel or Chub Mackerel		*Scomber colias*	Macarela	Maquereau blanc	12 14	
Mackerel, Atka		*Pleurogrammus monopterygius*	Lorcha de Atka	Maquereau d'Atka	23	
Mackerel, Atlantic		*Scomber scombrus*	Macarela o caballa del Atlántico	Maquereau	6 7 8 9	
Mackerel, Pacific Chub		*Scomber japonicus* and *Scomber spp.*	Macarela del Pacífico	Maquereau espagnol	18 19	
Mackerel, Jack; California (or Pacific) Jack Mackerel		*Trachurus symmetricus*	Jurel o charrito	Carangue symétrique	18 19 24	
Mackerel, King		*Scomberomorus cavalla*	Carito o peto	Thazard	14	
Mackerel, Spanish		*Scomberomorus maculatus*	Sierra	Thazard atlantique	14	
Mako, Shortfin	▲	*Isurus oxyrinchus*	Tiburón mako	Mako	18	
Marlins		*Makaira spp.*	Marlines	Marlins	10 18	
Marlin, Atlantic White; White Marlin	▲	*Tetrapturus albidus*	Marlín blanco o aguja blanca	Marlin blanc	9 10 12	
Marlin, Black		*Makaira indica*	Marlín negro	Makaire noir	17 18	
Marlin, Atlantic Blue		*Makaira nigricans*	Marlín azul del Atlántico	Makaire bleu de l'Atlantique	10	
Marlin, Indo-Pacific Blue		*Makaira mazara*	Marlín azul del Indo-Pacífico	Makaire bleu indo-pacifique	17 18	
Marlin, Striped		*Tetrapturus audax*	Marlín rayado	Marlin rayé	17 18	
Menhaden, Atlantic		*Brevoortia tyrannus*	Lacha del Atlántico o lacha	Alose tyran	11 13	
Mobula, Smoothtail	▲	*Mobula thurstoni*	Manta diablo o diablo de mar (mantarraya gigante)	Mante vampire	14 17 18 19	
Mobula, Spinetail	▲	*Mobula japanica*	Manta de espina o de aguijón (mantarraya gigante)	Mante aiguillat	17 19	
Mojarra, Pacific Flagfin		*Eucinostomus currani*	Mojarra tricolor	Blanche drapeau du Pacifique	17	
Mudsucker, Longjaw		*Gillichthys mirabilis*	Chupalodo grande	*Gillichthys mirabilis*†	18 19 20	
Mudsucker, Shortjaw		*Gillichthys seta*	Chupalodo chico	*Gillichthys seta*†	18 §	
Mullets		*Mugil spp.*	Lisas	Mulets	13 18	
Ocean Perch, Pacific		*Sebastes alutus*	Gallineta del Pacífico	Sébaste à longue mâchoire	20	
Pipefish, Opossum; Short-Tailed Pipefish	▲	*Microphis brachyurus*	Pez pipa culebra	Syngnathe à queue courte	12 13	
Pipefish, Texas	▲	*Syngnathus affinis*	Pez pipa texano	*Syngnathus affinis*†	13	

▲ species at risk   * introduced and invasive species   § endemic species   † species for which no common name could be found in the literature consulted.

Common Name in English (Some species have more than one common name)	At risk	Nom latin	Nombre común en español (Algunas especies tienen más de un nombre común)	Nom commun en français (certaines espèces ont plus d'un nom commun)	Featured in Ecoregion	Photo (page)
Plaice, American		*Hippoglossoides platessoides*	Platija americana	Plie canadienne	6	
Pollock; Boston blues; Coalfish		*Pollachius virens*	Carbonero	Goberge	6 7 9	
Pollock, Walleye; Alaskan Pollock; Pacific Pollock		*Theragra chalcogramma*	Abadejo (o colín) de Alaska	Goberge d'Alaska	1 2 22 23	
Pompano, Blackblotch		*Trachinotus kennedyi*	Pámpano o palometa	Pompaneau argenté	18	
Puffer, Bullseye		*Sphoeroides annulatus*	Botete diana	Compère Diane	17 18	
Ray, Atlantic Devil; Atlantic Manta		*Mobula hypostoma*	Manta del golfo (mantarraya gigante)	Mante diable	12 14 15	
Ray, Bat	▲	*Myliobatis californica*	Raya tecolote o murciélago	Aigle de mer técolette	18 19 20	
Ray, California Butterfly	▲	*Gymnura marmorata*	Raya mariposa californiana	Raie-papillon californienne	18	
Ray, Chilean Round; Blotched Stingray		*Urotrygon chilensis*	Raya pinta de espina	Raie ronde chilienne	17	
Ray, Giant Manta; Devil Fish	▲	*Manta birostris*	Manta gigante o voladora (mantarraya gigante)	Mante géante	14 17 18 19	*109*
Ray, Golden Cownose	▲	*Rhinoptera steindachneri*	Raya gavilán	Mourine du Pacifique	18	
Ray, Longnose (or Snouted) **Eagle**	▲	*Myliobatis longirostris*	Raya águila picuda	Aigle de mer espadon	18	
Ray, Pacific Manta		*Manta hamiltoni*	Mantarraya (gigante)	*Manta hamiltoni*†	14	
Ray, Pygmy Devil; Munk's Devil Ray	▲	*Mobula munkiana*	Manta violácea (mantarraya gigante)	Mante de Munk	17 18 19	
Ray, Sicklefin Devil; Box Ray, Chilean Devil Ray	▲	*Mobula tarapacana*	Manta cornuda (mantarraya gigante)	Mante chilienne	17 18 19	
Redfish		*Sebastes mentella*	Gallineta nórdica	Sébaste de l'Atlantique	6 7	
Rockfish, Blue		*Sebastes mystinus*	Rocote azul	Sébaste bleu	20 21 23	*115*
Rockfish, Bocaccio	▲	*Sebastes paucispinis*	Rocote bocaccio	Bocaccio	19 20 21 23	
Rockfish, Canary		*Sebastes pinniger*	Rocote canario	Sébaste canari	19 20 21	
Rockfish, Mexican		*Sebastes macdonaldi*	Rocote mexicano	Sébaste de corail	18 19 20	
Roosterfish		*Nematistius pectoralis*	Papagallo	Grand coq-de-mer	18	
Sablefish		*Anoplopoma fimbria*	Bacalao negro	Morue charbonnière	20 23	
Sailfish		*Istiophorus spp.*	Peces vela	Voiliers	10 17 18	
Salmon, Atlantic		*Salmo salar*	Salmón atlántico	Saumon atlantique	6 7 22 *	
Salmon, Chinook	▲	*Oncorhynchus tshawytscha*	Salmón real o chinook	Saumon quinnat	20 21 22 23	
Salmon, Chum	▲	*Oncorhynchus keta*	Salmón keta o chum	Saumon kéta	2 21 22 23	
Salmon, Coho	▲	*Oncorhynchus kisutch*	Salmón plateado o coho	Saumon coho	2 20 21 22 23	*8–9*
Salmon, Pink		*Oncorhynchus gorbuscha*	Salmón rosado	Saumon rose	1 2 20 21 22 23	*119*
Salmon, Sockeye		*Oncorhynchus nerka*	Salmón rojo	Saumon rouge ou sockeye	20 21 22 23	
Sand Bass, Goldspotted		*Paralabrax auroguttatus*	Cabrilla extranjera	Serran doré	18	
Sardine, Pacific; South American Pilchard		*Sardinops sagax*	Sardina monterrey	Pilchard du Chili	18 20	
Sargo; Xantic Sargo		*Anisotremus davidsonii*	Sargo rayado	Lippu du roche	18 19 20	
Sawfish (See Smalltooth or Largetooth)	▲	*Pristis spp.*	Peces sierra	Poissons-scies	12 13 14 17 18 19	
Sawfish, Largetooth	▲	*Pristis perotteti*	Pez sierra de diente largo	Poisson-scie grandent	13 14 17 18 19	
Sawfish, Smalltooth	▲	*Pristis pectinata*	Pez sierra peine	Requin-scie, poisson scie	12 13 14 17 18 19	
Scad, Bigeye		*Selar crumenophthalmus*	Jurel o chicharro ojón	Sélar	11 14	*142–143*
Scorpionfish, California		*Scorpaena guttata*	Lapón o escorpión californiano (pez roca)	Rascasse californienne	18 19 20	
Sculpin, Arctic		*Artediellus uncinatus*	Charrasco espinoso *Artediellus uncinatus*† (familia *Cottidae*)	Hameçon neigeux; crapaud de mer	3 4 5 6	

▲ species at risk　＊ introduced and invasive species　§ endemic species　† species for which no common name could be found in the literature consulted.

Common Name in English (Some species have more than one common name)	At risk	Nom latin	Nombre común en español (Algunas especies tienen más de un nombre común)	Nom commun en français (certaines espèces ont plus d'un nom commun)	Featured in Ecoregion	Photo (page)
Sea Bass, (Atlantic) Black		*Centropristis striata*	Serrano estriado o lubina negra	Bar noir	8 11	
Sea Bass, Giant		*Stereolepis gigas*	Mero pescada	Bar géant	18 19 20	
Sea Bass, White; White Weakfish		*Atractoscion nobilis*	Corvina blanca o cabaicucho	Acoupa blanc	18 19 20	
Seahorse, Dwarf	▲	*Hippocampus zosterae*	Caballito de mar enano	*Hippocampus zosterae* †	13	
Seahorse, Pacific		*Hippocampus ingens*	Caballito de mar del Pacífico	Hippocampe géant du Pacifique	18	
Sergeant Major, Panamic (damselfish)		*Abudefduf troschelii*	Chopa o sargento mayor	Demoiselle	17	
Shad, Alabama	▲	*Alosa alabamae*	Sábalo de Alabama	Alose de l'Alabama	13	
Shad, American; Atlantic Shad		*Alosa sapidissima*	Sábalo americano	Alose savoureuse	8 11	
Shad, Threadfin		*Dorosoma petenense*	Topote	Alose fil	18 *	
Shark, Atlantic Sharpnose	▲	*Rhizoprionodon terraenovae*	Cazón de ley	Requin à nez pointu de l'Atlantique	14 15 18	
Shark, Basking	▲	*Cetorhinus maximus*	Tiburón peregrino	Pèlerin	14 19	
Shark, Blacktip	▲	*Carcharhinus limbatus*	Tiburón volador o puntas negras	Requin bordé	14 15 18	
Shark, Broadnose Sevengill	▲	*Notorynchus cepedianus*	Tiburón de 7 branquias	Platnez	18	
Shark, Bull	▲	*Carcharhinus leucas*	Tiburón chato	Requin bouledogue	14 15	
Shark, Dusky	▲	*Carcharhinus obscurus*	Tiburón oscuro o arenero	Requin obscur	8 11 12 13 18	
Shark, Great White	▲	*Carcharodon carcharias*	Tiburón blanco	Grand requin blanc	14 17 18 19	
Sharks, Hammerhead	▲	*Sphyrna spp.*	Tiburones martillo o cornudas	Requins-marteaux	16 17 18 19	
Shark, Lemon	▲	*Negaprion brevirostris*	Tiburón limón	Requin citron	18	
Shark, Leopard		*Triakis semifasciata*	Tiburón leopardo	Requin léopard	18 19 20	
Shark, Night	▲	*Carcharhinus signatus*	Tiburón nocturno u ojo verde	Requin de nuit	8 11 12 13	
Shark, Pacific Angel	▲	*Squatina californica*	Tiburón ángel o angelito	Ange de mer du Pacifique	18	
Shark, Pacific Sharpnose	▲	*Rhizoprionodon longurio*	Cazón bironche	Requin à nez pointu du Pacifique	18	
Shark, Sand Tiger; Grey Nurse Shark	▲	*Carcharias taurus*	Tiburón toro	Requin taureau	7 8 11 12 13	
Shark, Scalloped Hammerhead	▲	*Sphyrna lewini*	Cornuda común o tiburón martillo	Requin-marteau halicorne	16 17 18 19	
Shark, Silky	▲	*Carcharhinus falciformis*	Tiburón sedoso o tiburón piloto	Requin soyeux	14 15 16 17 18 19	
Shark, Smooth Hammerhead	▲	*Sphyrna zygaena*	Cornuda prieta o cruz	Requin-marteau commun	16 17 18 19	
Shark, Whale	▲	*Rhincodon typus*	Tiburón ballena	Requin-baleine	14 17 18 19	
Shark, Whitetip reef		*Triaenodon obesus*	Tiburón punta blanca de arrecife o cazón coralero	Requin corail	18 §	*102*
Sheephead, California		*Semicossyphus pulcher*	Vieja californiana	Labre californien	18 19 20	
Sierra, Gulf; Monterey Spanish Mackerel		*Scomberomorus concolor*	Sierra del golfo	Thazard de Monterey	18	
Sierra, Pacific (Gulf)		*Scomberomorus sierra*	Sierra (del Pacífico)	Thazard sierra	18	
Silverside, Delta		*Colpichthys hubbsi*	Pejerrey delta	Athérine delta	18 §	
Skate, Barndoor	▲	*Dipturus laevis*	Raya manchada americana	Grande raie	7 8 11	
Skate, California		*Raja inornata*	Raya de California	Raie de Californie	20	
Skate, Cortez		*Raja cortezensis*	Raya de Cortés	Raie de Cortez	18 §	
Skipjack, Black		*Euthynnus lineatus*	Bacoreta negra	Thonine	18	
Smelt, Delta	▲	*Hypomesus transpacificus*	Eperlano del delta	Éperlan du delta	20 §	
Snailfish		Family: *Liparidae*	Peces caracol	Limaces de mer	3 4	

▲ species at risk   * introduced and invasive species   § endemic species   † species for which no common name could be found in the literature consulted.

Common Name in English (Some species have more than one common name)	At risk	Nom latin	Nombre común en español (Algunas especies tienen más de un nombre común)	Nom commun en français (certaines espèces ont plus d'un nom commun)	Featured in Ecoregion	Photo (page)
Snappers	▲	*Lutjanus spp.*	Pargos y huachinangos	Vivaneau	12 14 18	
Snapper, Amarillo		*Lutjanus argentiventris*	Pargo amarillo o almazán	Vivaneau jaune	18	
Snapper, Colorado		*Lutjanus colorado*	Pargo colorado o listoncillo	Vivaneau amarante	18	
Snapper, Northern Red		*Lutjanus campechanus*	Huachinango del golfo	Vivaneau rouge	12 14	
Snapper, Pacific Dog		*Lutjanus novemfasciatus*	Pargo prieto o mulato	Vivaneau charbonnier	18	
Snapper, Pacific Red		*Lutjanus peru*	Huachinango del Pacífico	Vivaneau garance	18	
Snapper, Spotted Rose		*Lutjanus guttatus*	Pargo lunarejo o flamenco	Vivaneau rose	18	
Snapper, Whipper		*Lutjanus jordani*	Pargo colmillón	Vivaneau huachinango	18	
Snooks	▲	*Centropomus spp.*	Robalos	Brochets de mer	18	
Sole, Fantail		*Xystreurys liolepis*	Lenguado cola de abanico	Rite éventail	18 19 20	
Steelhead (Trout) or Rainbow Trout	▲	*Oncorhynchus mykiss*	Trucha arco iris	Truite arc-en-ciel	20 21 22	
Stingray, Cortez; Spotted Round Stingray		*Urobatis maculatus*	Raya lija de espina	Raie ronde tachetée	18 §	
Stingray, Diamond	▲	*Dasyatis dipterura*	Raya diamante	Pastenague à deux queues	18	
Sturgeon, Atlantic	▲	*Acipenser oxyrinchus*	Esturión del Atlántico	Esturgeon noir	7 8 11	
Sturgeon, Gulf	▲	*Acipenser oxyrinchus desotoi*	Esturión del golfo	*Acipenser oxyrinchus desotoi* †	13	
Sturgeon, Shortnose	▲	*Acipenser brevirostrum*	Esturión chato	Esturgeon à museau court	7 8 11	
Surfperch, Pink; Pink Seaperch		*Zalembius rosaceus*	Mojarra rosada o perca	*Zalembius rosaceus* †	18 19 20	
Surgeonfish, Razor		*Prionurus laticlavius*	Cirujano barbero o cochinito	Chirurgien barbier	17	93
Swordfish		*Xiphias gladius*	Pez espada	Espadon	10 17	
Thread Herring		*Opisthonema spp.*	Sardinas crinudas o arenques de hebra	Chardins	18	
Thread Herring, Deepbody		*Opisthonema libertate*	Sardina crinuda del Pacífico	Chardin du Pacifique	18	
Thread Herring, Middling		*Opisthonema medirastre*	Sardina crinuda machete	Chardin fil entrefin	18	
Thresher, Bigeye	▲	*Alopias superciliosus*	Tiburón grillo u ojón	Requin-renard à gros yeux	17 18 19	
Thresher, Pelagic	▲	*Alopias pelagicus*	Tiburón zorro o coludo	Requin-renard pélagique	17 18 19	
Tilefish		Family: *Malacanthidae*	Blanquillos, piernas o peces conejo	Tiles	10	
Tilefish, Great Northern	▲	*Lopholatilus chamaeleonticeps*	Conejo amarillo o corvinato	Achigan de mer	9 10	
Topminnow, Saltmarsh	▲	*Fundulus jenkinsi*	Sardinilla del Bravo	*Fundulus jenkinsi* †	12 § 13	
Totoaba	▲	*Totoaba macdonaldi*	Totoaba	Totoaba	18 §	
Triggerfish		Family: *Balistidae*	Peces ballesta	Balistes	14 18	
Triggerfish, Finescale		*Balistes polylepis*	Cochi, cochito o pez puerco	Baliste coche	18	
Triggerfish, Orangeside		*Sufflamen verres*	Cochito naranja o puerco naranja	Baliste calafate	18	
Tuna, Bigeye		*Thunnus obesus*	Atún ojo grande o patudo	Thon ventru	8 9 10 17 18	
Tuna, Bluefin		*Thunnus thynnus*	Atún aleta azul	Thon rouge	7 8 9 10 14 17 18	
Tuna, Pacific Bluefin		*Thunnus thynnus orientalis*	Atún aleta azul del Pacífico	Thon rouge du Pacifique	18	
Tuna, Skipjack		*Katsuwonus pelamis*	Barrilete listado	Bonite à ventre rayé	16 18	
Tuna, Yellowfin		*Thunnus albacares*	Atún aleta amarilla	Thon à nageoires jaunes	10 11 12 14 16 17 18	
Turbot, Diamond		*Hypsopsetta guttulata*	Platija diamante	Flet de diamant	18 19 20	
Varden, Dolly		*Salvelinus malma*	Salvelino	Dolly Varden	22	

▲ species at risk ∗ introduced and invasive species § endemic species † species for which no common name could be found in the literature consulted.

Common Name in English (Some species have more than one common name)	At risk	Nom latin	Nombre común en español (Algunas especies tienen más de un nombre común)	Nom commun en français (certaines espèces ont plus d'un nom commun)	Featured in Ecoregion	Photo (page)
Wahoo		Acanthocybium solandri	Peto o wahoo	Thazard bâtard	10 18	
Whitefish; Cisco		Coregonus spp.	Cisco, pez blanco	Corégones	2	
Wrasses		Family: Labridae	Señoritas y viejas	Vielles	10 17 18	
Wrasse, Rock		Halichoeres semicinctus	Señorita piedrera	Donzelle de roche	18 * 19 20	
Wrasse, Sunset		Thalassoma grammaticum	Señorita o vieja isleña	Girelle crépuscule	17	
Wreckfish		Polyprion americanus	Cherna	Cernier de l'Amérique	10	
Marine Invertebrates			**Invertebrados marinos**	**Invertébrés marins**		
Abalone, Black	▲	Haliotis cracherodii	Abulón negro	Ormeau noir	19 20 21	
Abalone, Green	▲	Haliotis fulgens	Abulón verde	Ormeau vert	19 §	
Abalone, Pink	▲	Haliotis corrugata	Abulón amarillo	Ormeau rose	19 §	
Abalone, Pinto or Northern	▲	Haliotis kamtschatkana	Abulón del norte o perlado	Ormeau nordique	20 21	
Abalone, White	▲	Haliotis sorenseni	Abulón blanco	Ormeau blanc	19 §	
Anemones		Anemone spp.	Anémonas	Anémones	3 4	122
Brachiopod, Inarticulated	▲	Lingula reevii	Braquiópodo inarticulado Lingula reevii †	Brachiopode inarticulé	24	
Bursa gastropod mollusks		Bursa spp.	Bursa spp. (moluscos gasterópodos)	Bursa spp. (molusques gastéropodes)	17	
Clam, Asian		Potamocorbula amurensis	Almeja china	Palourde d'Asie	20 * 21 *	
Clam, Purple Varnish		Nuttallia obscurata	Almeja Nuttallia obscurata	Nutallie obscure	21 *	
Conch, Purpura	▲	Plicopurpura pansa	Caracol púrpura o caracol de tinta	Plicopurpura pansa †	16 17	
Conch, Queen or Pink	▲	Strombus gigas	Caracol reina o rosado	Strombe géant	15	
Coral, Bird's Nest		Pocillopora damicornis	Coral risco (coral hermatípico)	Pocillopora damicornis †	17	
Corals, Black	▲	Family: Antipathidae	Corales negros	Coraux noirs	15	
Corals, Bubblegum		Family: Paragorgiidae	Corales gorgonia	Corails « bubblegum »; corails Paragorgia arborea †	22 23	
Coral, Cauliflower		Pocillopora verrucosa	Coral risco (coral hermatípico)	Corail verruqueux	17	
Corals, Dendrophelia		Dendrophelia spp.	Dendrophelia spp. † (coral)	Dendrophelia spp. †	10	
Coral, Elkhorn	▲	Acropora palmata	Coral cuerno de alce	Corail corne d'élan	12 13 15	
Corals, Enallopsammia; Stony Corals		Enallopsammia spp.	Enallopsammia spp. † (coral)	Enallopsammia spp. †	10	
Corals, Gorgonian		Family: Gorgoniidae	Corales gorgonia	Gorgones	20 22	
Coral, Hawaiian Rice	▲	Montipora dilatata	Coral hawaiano Montipora dilatata †	Montipora dilatata †	24	
Coral, Ivory Tree; Oculina Coral	▲	Oculina varicosa	Coral oculina o arbusto de marfil	Oculina varicosa †	10 11	
Coral, Lophelia; Stony Coral		Lophelia pertusa	Lophelia pertusa † (coral)	Lophelia pertusa †	10 13	58
Coral, Red-Tree		Primnoa resedaeformis	Primnoa resedaeformis † (Coral rojo)	Primnoa resedaeformis †	23	
Coral, Snowflake; Branched Pipe Coral		Carijoa riisei	Coral copo de nieve	Telesto blanc	24 *	
Coral, Staghorn	▲	Acropora cervicornis	Coral cuerno de ciervo	Corail en cornes de cerf	12 13 15	
Corophium (amphipod, mud shrimp)		Corophium spp.	Corophium spp. † (crustáceos anfípodos)	Corophium spp.	7	
Crab, Blue		Callinectes sapidus	Jaiba azul	Crabe bleu	8 11 12 18	
Crab, Chinese Mitten		Eriocheir sinensis	Cangrejo chino con mitones	Crabe chinois	20 *	
Crab, Dungeness		Cancer magister	Cangrejo dungeness	Crabe dormeur	20	
Crab, European Green		Carcinus maenas	Cangrejo verde europeo o común	Crabe européen	20 * 21 * 22 *	

▲ species at risk   * introduced and invasive species   § endemic species   † species for which no common name could be found in the literature consulted.

Common Name in English (Some species have more than one common name)	At risk	Nom latin	Nombre común en español (Algunas especies tienen más de un nombre común)	Nom commun en français (certaines espèces ont plus d'un nom commun)	Featured in Ecoregion	Photo (page)
Crab, Horseshoe		*Limulus polyphemus*	Cangrejo herradura o cacerola	Limule	8	62
Crabs, King		*Paralithodes spp., Lithodes aequispinus*	Cangrejo real	Crabe royal	1	
Crab, Snow		*Chionoecetes opilio*	Cangrejo de nieve	Crabe des neiges	7	
Crayfish (incl. Texas 'Red Swamp' Crayfish)		*Procambarus spp.*	Acociles, Cangrejos de río (incl. cangrejo rojo de marjal)	Écrevisses (incl. Écrevisse rouge des marais)	13	
Ficus gastropod mollusks		*Ficus spp.*	Caracol higo	*Ficus spp.*	17	
Jellyfish, Big Pink		*Drymonema dalmatinum*	Medusa rosada	Méduse irradiante	15 *	
Krill		*Euphausia spp.*	Krill	Krills	1	
Lobster, American		*Homarus americanus*	Langosta americana	Homard d'Amérique	7 8	
Lobster, Caribbean Spiny		*Panulirus argus*	Langosta del Caribe	Langouste blanche	12 15	81
Lobsters, Spiny		*Panulirus spp.*	Langostas	Langoustes	11	
Lobster, Squat		*Eumunida picta*	Cangrejo de aguas profundas	Galatée rouge	10	58
Octocoral, Northern		*Paragorgia arborea*	Coral gorgonia	*Paragorgia arborea*†	22 23	
Octocorals, Primnoidae		Family: *Primnoidae*	Coral gorgonia	Octocorail Primnoidae	22	
Octopus, Maya		*Octopus mayaes*	Pulpo maya	*Octopus mayaes*†	14 §	
Oyster, Eastern		*Crassostrea virginica*	Ostión americano u ostión del Este	Huître américaine ou huitre de l'est	8 11 13	
Oyster, Pacific Giant		*Crassostrea gigas*	Ostra gigante del Pacífico	Huître creuse du Pacifique	18 * 21 *	
Rangia, Atlantic (Brackish Water Clam)		*Rangia cuneata*	Rancia americana, Almeja de agua salobre o gallito	Palourde d'eaux saumâtres, palourde de l'Atlantique	14	
Scallops		Family: *Pectinidae*	Vieiras o escalopas	Pétoncles	7 22	
Sea Cucumbers		Family: *Holothuroide*	Pepinos de mar	Holothuries; concombres de mer	2 18	
Sea Squirt; Ascidian; Colonial Tunicate; Compound Sea Squirt		*Didemnum vexillum*	Ascidia *Didemnum vexillum*† (organismo tunicado)	*Didemnum vexillum*†	7 * 8 * 20 * 21 *	
Sea Stars (generic)		Phylum *Echinodermata*, Class *Asteroidea*	Estrellas de mar	Étoiles de mer	2 3 4 7 21	122
Sea Star, Ochre		*Pisaster ochraceus*	*Pisaster ochraceus*†	Étoile ocrée	21	122
Sea Urchins, Long-spined		*Diadema spp.*	Erizo diadema	Oursins diadèmes	2 20	
Shawl, Spanish		*Flabellina iodinea*	Nudibranquio morado	*Flabellina iodinea*†	19	100–101
Shells, Harp		*Harpa spp.*	Harpas	*Harpa spp.*	17	
Shells, Murex		*Hexaplex spp.*	Caracol chino, murex	Murex	17 18	
Shell, Spindle		*Fusinus spp.*	Caracol chilillo o chile	*Fusinus spp.* †	17	
Shrimp, Blue		*Litopenaeus stylirostris*	Camarón azul	Crevette bleue	18	
Shrimp, Brown		*Penaeus aztecus*	Camarón café	Crevette brune	13 14 18	
Shrimp, Northern; Maine Shrimp		*Pandalus borealis*	Camarón boreal	Crevette nordique	7	
Shrimp, Pink	▲	*Farfantepenaeus duorarum*	Camarón rosado	Crevette rose	12 13 14 15	
Shrimp, Royal Red		*Pleoticus robustus*	Camarón rojo real	Salicoque royale rouge	12 13 14 15	
Shrimp, Whiteleg; Pacific White Shrimp		*Litopenaeus vannamei*	Camarón blanco	Crevette pattes blanches	13 14 * 15 * 18	
Snails, Cantharus		*Cantharus spp.*	Cambutes (moluscos gasterópodos)	Cantharus	17	
Squid, California Market		*Loligo opalescens*	Calamar de California	Calmar opale	20	
Squid, Jumbo; Humboldt Squid		*Dosidicus gigas*	Calamar gigante	Encornet géant	18	
Stickyhydroid		*Eudendrium ramosum*	Hidrozoario *Eudendrium ramosum*†	*Eudendrium ramosum*†	19	
Whelk, Veined Rapa		*Rapana venosa*	Busano veteado	Rapana veiné	8 *	

▲ species at risk   * introduced and invasive species   § endemic species   † species for which no common name could be found in the literature consulted.

Common Name in English (Some species have more than one common name)	At risk	Nom latin	Nombre común en español (Algunas especies tienen más de un nombre común)	Nom commun en français (certaines espèces ont plus d'un nom commun)	Featured in Ecoregion	Photo (page)
Vegetation: Mangroves, Aquatic Plants and Algae			**Vegetación: mangles, plantas acuáticas y algas**	**Végétation : Mangroves, plantes aquatiques et algues**		
Algae, Brown; Sargassum		Class: *Phaeophyceae*	Sargazo (alga parda o café)	Sargasse japonaise (algue brune)	7 16* 17*	
Algae, Ice		*Melosira arctica*	Alga de hielo	Algue glaciale	3 4	
Barley, Wild		*Hordeum jubatum*	Cebada silvestre	Orge queue-d'écureuil	7	
Codium (Green Alga)		*Codium oaxacensis*	*Codium oaxacensis*† (alga verde)	*Codium oaxacensis*†	16§	
Cordgrass, Saltmarsh or Atlantic		*Spartina alterniflora*	Espartina del Atlántico	Spartine alterniflore	7 13 21*	
Cordgrass, Saltmeadow		*Spartina patens*	Espartina o hierba de sal	Spartine étalée	7 13	
Cypress, Bald		*Taxodium distichum*	Ciprés calvo o de los pantanos	Cyprès chauve	13	
Eelgrass		*Vallisneria spp.*	Vallisneria (pasto marino)	Zostères	21 22	
Eelgrass, Japanese		*Zostera japonica*	Pasto marino *Zostera japonica*†	Zostère asiatique	21*	
Gracilaria salicornia (Red Alga)		*Gracilaria salicornia*	*Gracilaria salicornia*† (alga roja)	*Gracilaria salicornia*†	24*	
Hyacinth, Water		*Eichhornia crassipes*	Jacinto de agua	Jacinthe d'eau	12*	
Hypnea musciformis (Red Alga)		*Hypnea musciformis*	*Hypnea musciformis*† (alga roja)	*Hypnea musciformis*†	24*	
Ishige foliacea (Red Alga)		*Ishige foliacea*	*Ishige foliacea*† (alga roja)	*Ishige foliacea*†	18*	
Jelly, Comb; Ctenophore		*Pleurobrachia bachei* and *Beroe spp.*	Farolito de mar (ctenóforo)	Groseille de mer	18	
Kelp, Bull; "Mermaid's Bladder"		*Nereocystis luetkeana*	Kelp cabeza de toro (alga parda gigante)	Nereocystis de Lutke	20 21 22 23	
Kelp, Giant		*Macrocystis pyrifera; M. integrifolia*	Kelp, sargazo gigante	Algue géante	7 19 20 21 22	
Knotweed, Japanese		*Fallopia japonica* o *Polygonum cuspidatum*	Falopia japonesa	Renouée du Japon	22*	
Loosestrife, Purple		*Lythrum salicaria*	Salicaria	Salicaire pourpre	7* 12* 22*	
Manateegrass		*Syringodium filiforme*	Pasto de manatí	Herbe à lamantin	12 13	
Mangrove, Black		*Avicennia germinans*	Mangle negro	Palétuvier noir ou bois de mèche	12 13 15 16 17	*160–161*
Mangrove, Button		*Conocarpus erectus*	Mangle de botoncillo	Palétuvier gris	12 13 15 16 17	
Mangrove, Dwarf; Grey Mangrove		*Avicennia marina*	Mangle enano o gris	Palétuvier nain	13	
Mangrove, Oriental		*Bruguiera sexangula*	Mangle oriental	Palétuvier oriental	24*	
Mangrove, Red		*Rhizophora mangle*	Mangle rojo	Palétuvier rouge	12 13 15 16 17 24*	*88–89, 160–161*
Mangrove, White		*Laguncularia racemosa*	Mangle blanco	Palétuvier blanc	12 13 15 16	*160–161*
Pondweed, Curly-leaved		*Potamogeton crispus*	Potamogeton de hoja rizada	Potamot crépu	12*	
Rush, Needlegrass		*Juncus roemerianus*	Junco pasto de aguja	*Juncus roemerianus*†	13	
Sea Lavender		*Limonium spp.*	Estática o lavanda de mar	Lavandes de mer	7	
Sea Plantain		*Plantago maritima*	Plantago o llantén de mar	Plantain de mer	7	
Seagrass, Johnson's	▲	*Halophila johnsonii*	Pasto marino de Johnson	*Halophila johnsonii*	12 13	
Shoalweed		*Diplanthera wrightii*	*Diplanthera wrightii*† (pasto marino)	*Diplanthera wrightii*†	12 13	
Spikegrass, Marsh		*Distichlis spicata*	Zacate salado	Distichlis en épi	7	
Torpedograss		*Panicum repens*	Pasto torpedo	Panic rampant	12*	
Turtlegrass		*Thalassia testudinum*	Pasto de tortuga	*Thalassia testudinum*†	12 13	
Widgeongrass		*Ruppia maritime*	Broza fina	Ruppie	12 13	

▲ species at risk   * introduced and invasive species   § endemic species   † species for which no common name could be found in the literature consulted.

Semipalmated sandpipers at the Bay of Fundy, New Brunswick. *Photo:* Fred Bruemmer/DRK PHOTO